APPLIED AERONAUTICS

— THE AIRPLANE —

First Edition
1918

Published by

Airplane Engineering Department
McCook Field, Dayton, Ohio
U. S. A.

PREFACE

TODAY the subject of aeronautics appeals to nearly everyone, due largely to the wonderful progress which has been made in the development and utility of the airplane since the beginning of the war.

Although the airplane is among the most important features of modern warfare, yet there are comparatively few comprehensive texts which the beginner in aviation can readily understand. In many cases the student aviator and the aero mechanic have had but a limited technical training nor have many of them been specially trained in mathematics, particularly in the higher branches. Therefore, many of the present-day text-books on aeronautics, which the aeronautical engineer would consider quite elementary, are very difficult for the beginner.

Realizing the difficulty under which the beginner labors, the Airplane Engineering Department felt that an elementary course in applied aeronautics and practical aviation was needed, and has brought forth this volume with the hope of filling this need, primarily for the instruction of the men at McCook Field. This book, however, is not intended as a text on the design of airplanes, but was written solely with the thought of imparting a clear idea of the principles of flight, together with such other practical information as can be applied readily by the man in the shop or in the air.

The text is based primarily upon lectures given at the U. S. Army School of Military Aeronautics, Ohio State University. The Technical Publications Department at McCook Field, however, has rearranged, reclassified and rewritten a great deal of the text matter contained in these lectures in order to make the book as clear and instructive as possible. Some of the lectures as given by the instructors have been split up into several chapters, and, where it was felt that certain parts of these lectures could be enlarged upon, the Technical Publications Department has taken the liberty of making such changes as were thought advisable. A number of illustrations and diagrams have been added to those used by the instructors.

The Airplane Engineering Department therefore gratefully acknowledges its obligations to Mr. H. C. Lord for various sections in the first chapter on the theory of flight, and jointly to Mr. Lord and Mr. G. T. Stankard for a large part of the chapters on instruments. It also wishes to express its thanks to Mr. W. A. Knight for the information furnished by him on rigging and alignment of planes. These men are instructors in the U. S. Army School of Military Aeronautics, Ohio State University.

It is hoped that this book will assist the student aviator and the aero mechanic in getting one step nearer their goals as efficient and experienced units in the U. S. Air Service, and if it accomplishes this result the Airplane Engineering Department will feel amply repaid for its efforts in bringing it forth.

Dayton, Ohio, U. S. A., July 1, 1918.

CONTENTS

Contents

APPLIED AERONAUTICS

— THE AIRPLANE —

Chapter 1

THEORY OF FLIGHT

Investigating wind action—Constant values—Studying action of wind—Streamline shapes—Head resistance—Lift, drift and angle of attack—Suction on top of plane—Center of pressure—Cambered planes—Horizontal flight—Engine power—Power to climb—Stability.

IN THIS age of mechanical flight everyone is interested in airplanes. But very few people, however, clearly grasp the underlying principles. The theory involved, nevertheless, may be demonstrated by simple experiments, so that the reader with only an elementary knowledge of mathematics and mechanics can understand.

The simplest principle of airplane flight may be demonstrated by plunging the hand in water and trying to move it horizontally, after first slightly inclining the palm so as to meet, or attack, the fluid at a small angle. It will be noticed at once that although the hand remains very nearly horizontal, and though it is moved horizontally, the water exerts upon it a certain amount of pressure directed nearly vertically upwards and tending to lift the hand. This is a fair analogy to the principle underlying the flight of an airplane. The wings of the plane are set at a small angle, and the plane is pushed or pulled through the air by the propeller, which receives its power from the engine. The action of the air on the wings, inclined at an angle, tends to lift the plane just as the action of the water on the hand, inclined at a small angle, has a tendency to raise the hand out of the water.

Investigating Wind Action

A rough form of apparatus for studying laws of wind resistance is shown in Fig. 1. The arm E hinged at C carries a rectangular plane B. The adjustable weight D, supported by the arm F, is used to balance the pressure of the wind from the blower A. The pressure exerted on the plane B can then be measured by moving the weight D along the arm F until B floats with the wind.

Professor Langley, in another experiment, proved that we can investigate the action of the wind upon various forms of surfaces as well by directing a current of air of known velocity against the surface held in position, and weighing the reactions, as we can by forcing the plane itself through still air. The special apparatus used was mounted on the end of a revolving arm driven by a steam engine as is shown in Fig 2. The chronograph, a recording instrument, was used to measure the velocity or number of revolutions of the table in a given time.

By such a method as that shown in Fig. 1, and that of Professor Langley, it is easy to see that the laws of pressure and velocity can be determined readily. Methods such as these have been-used in determining that *the wind resistance varies as the square of the velocity.*

In other words, if the velocity is doubled it follows that the resistance is increased four times, or if velocity is five times as great, the wind resistance is twenty-five times as large.

Constant Values

It would therefore seem to need no experimentation to prove that if we increase the surface B (Fig. 1) we would increase the pressure in direct proportion to the increase in surface area. Now if we were to increase both the velocity and the area of surface, we would increase the pressure proportionally to the product of the square of the velocity and the area of the surface. Thus if we were to raise the velocity of the air three times, the resistance would be increased nine times, and if we then doubled the surface we would double the resistance, which has already been increased nine times, making a total increase of eighteen-fold.

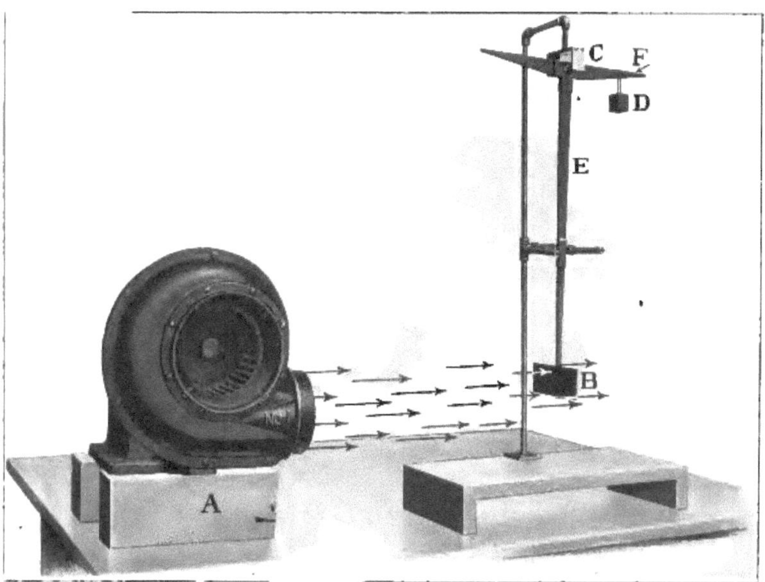

Fig. 1—Elementary apparatus for studying laws of wind resistance

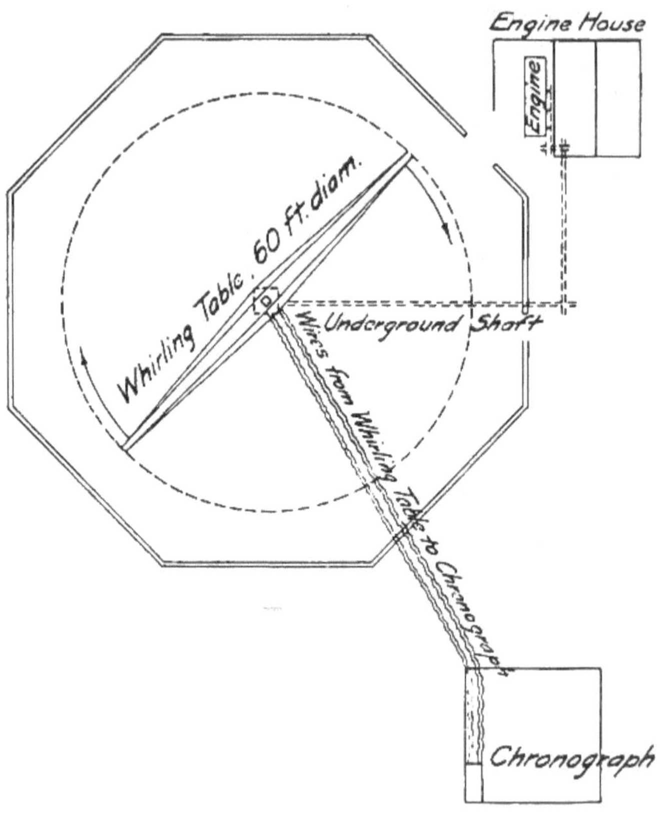

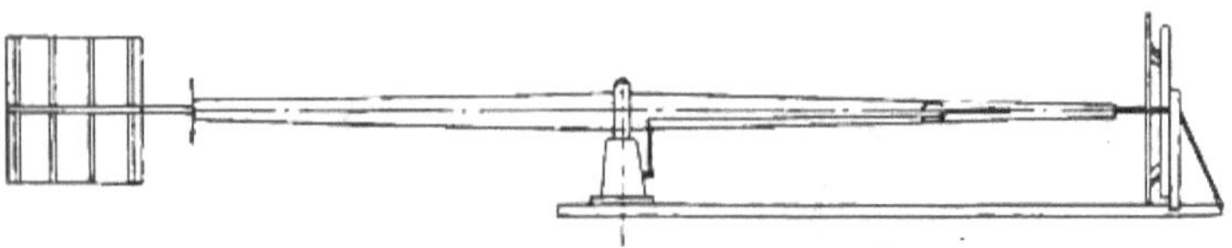

Side Elevation of Whirling Arm

Fig. 2—Prof. Langley's apparatus for investigating wind action on various forms of surfaces

There is still another factor to take into consideration in calculating wind pressures, and that is the shape of the surface. To take that into account we must use what is called a *constant*, the value of which is determined by experiments for each particular shape of surface.

The following explanation will enable one to see very clearly the meaning of the term constant and how its value is determined. First let us explain the term *formula* which is merely a sentence tersely expressed. To attempt to make a study of flight without formulæ would make it necessary to express relations between quantities in long paragraphs of words that could more readily be stated in simple equations. Thus if it were desired to state the rule that the quantity A multiplied by twice the quantity B is equal to C, the formula representing this would be:

$$A \times 2B = C$$

Each letter or symbol in a formula represents some factor that is substituted when its value is known. If $A = 16$, and $B = 4$, then $C = 128$, since the rule interpreted reads: $16 \times 8 = 128$.

Derived and empirical equations.—The term equation simply means that the quantities on one side of the equal sign are equivalent or equal to the quantities on the other side. Equations are of two kinds, derived and empirical. A derived equation is susceptible to proof, by use of mathematical processes. An empirical equation is neither derived nor proven. It is merely a statement of the results of experiment regardless of mathematical proof.

In many branches of engineering, empirical formulæ are constantly used, and in aviation the lack of a satisfactory basic theory of air flow makes empirical formulæ based on experiment most necessary. Empirical formulæ are really experimental averages.

The term *constant* can now be fully explained and it will be seen how beautifully it works out in a formula. It is often found necessary,

Fig. 3—Illustrating meaning of term "projected area"

especially in an experimental field, to introduce numerical constants to balance the two sides of an equation. For example, the pressure on a surface, as we have already learned, is equal to a constant times the projected area of the surface (see Fig. 3) times velocity squared, or expressing the same quantities in a formula,

$$P = KSV^2$$

where P = Pressure, K = Constant, S = Projected surface area, V^2 = Velocity squared

The exact value of the constant K for any surface is determined experimentally by wind tunnel tests. So valuable have wind tunnels proven for such determinations that several of the large airplane builders now have installed them in their plants.

In solving a problem it might be known that the pressure P varies as the area of the surface and the velocity squared, but we could not express this relation in an equation capable of solution until a numerical value for K is determined for the particular shape subjected to the wind pressure, such as the shape illustrated in Fig. 3. Each different shape of surface requires a different value for K, which can be determined experimentally.

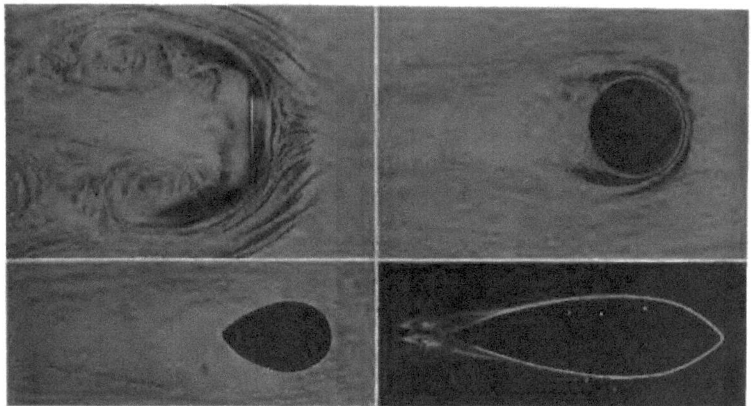

From Loening's *Military Aeroplanes*

Fig. 4—Showing forms of eddies set up by various shaped surfaces in air currents

The majority of formulæ for air pressures involve constants, and the great advance in designing during the past two years may be traced directly to the determinations by the aerodynamic laboratories, of better values of these constants, for use in empirical formulæ. So when M. Eiffel, or other men of authority, inform us that the constant K for a flat shape is .003, we accept the value just as we do the report of a chemist who tells us the composition of an alloy.

Studying Action of Wind

A study will now be made of the action of the wind upon various surfaces. Fig. 4 shows what would be seen if the air were filled with smoke or other particles and blown from the blower in Fig. 1 past the surface, and then an instantaneous photograph made. You will note how in the first picture the air is piled up in front of the surface and how it eddies and whirls behind it, thus showing that the disturbance extends far beyond the actual obstruction itself. Note the decrease in the eddies in the cases of the other shapes in Fig. 4.

The sphere offers less obstruction than the flat plane, and the two peculiar shapes at the bottom offer still less. Such bodies are said to be *streamline*. The big end is the advancing end. An old rule for the design of a whaling ship was that "it should have the head of a cod and the tail of a mackerel." With such streamline shapes, K is much smaller in value than for a flat square plane.

Parasite Resistance

A picture of a typical airplane is shown in Fig. 6. Notice that all the struts, wires, landing wheels and the fuselage or body offer resistance to passage through the air—a resistance which must be overcome by the engine. The sum total of the separate resistances of all these parts is called the *parasite resistance*. This wastes power and so all such parts are carefully *streamlined* wherever possible.

Fig. 5—Experiment showing lift of inclined surface in air current

Note the wings or aerofoils, two on each side, one above and one below, and at the rear a vertical rudder R in front of which is a vertical fin V, and the horizontal fin H, the back part of which can be turned up or down by the pilot. The effect of this is to cause the machine to point up or down and thus change the angle at which the relative wind strikes the aerofoils. This change, as we will see, has much to do with the flying of the machine.

Lift, Drift and Angle of Attack

Thus far we have found a lot of things about an airplane which would tend to prevent its flying. Now let us study Fig. 5. Here we have a plane B fastened so that it makes a small angle with the direction of the wind from the blower A. The arm is hinged at C, and balanced by the weight D, so that when the movable weight W is pushed back to C the plane B will be slightly too heavy. When the blower A is started

Fig. 6—Curtiss airplane, showing control surfaces

the plane B instantly lifts and the amount of this lift may be measured by the movable weight W. If we replaced this model by one exactly like it except that the plane B makes a much smaller angle with the relative wind we would find that the movable weight W would have to be much nearer C than before. This simple experiment proves the existence of a force which tends to lift the plane and further that this force is greater as the angle is increased. This angle is called the *angle of attack* that the plane B makes with the air stream. The force which tends to raise the plane is called the *lift*, and evidently its value must depend upon the profile of the plane, the velocity squared, and the angle of attack.

Besides the lift, there is another force which is due to the plane's velocity through the air, called the *drift*. This force is due to the fact that the plane itself offers resistance to forward motion through the air. In Fig. 7, A represents a bubble of air, BC a plane moving in the direction of the arrows. Now evidently one of two things must happen. Either the plane must force the bubble of air down or the bubble of air must force the plane up. This resistance that the bubble of air offers to being displaced, as we have seen, depends upon the square of the velocity with which it is forced out of the way. The total resistance offered by the bubble to the movement of the plane may be represented by the force P acting at right angles to the surface of the plane. The horizontal and vertical components of P are represented by D and L, respectively.

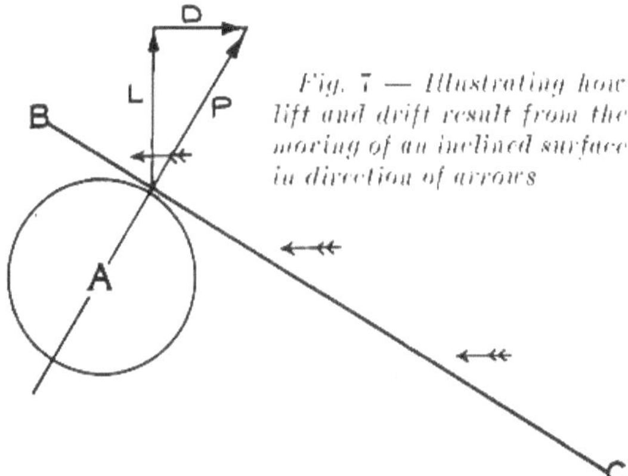

Fig. 7 — Illustrating how lift and drift result from the moving of an inclined surface in direction of arrows

If we were to let the air force on the surface have its way, it would push the surface upwards in the direction of L and backwards in the direction of D at the same time.

So we put weight on the surface, enough to overcome the force L, and then quite logically call this force the lift. And for D, we push against it, with the thrust from a propeller, and we call D the drift.

This simple explanation enables us at once to state the reason why flight in heavier-than-air machines is possible. By pushing the inclined surface into the air with a horizontal force D, we create a pressure on the surface equal to P, the force of which D is the horizontal component. But by doing so we have also created the other component L, which is a lifting force, capable of carrying weights into the air.

Consideration of this resolution into lift and drift at once indicates that the characteristics to be sought for in a surface are great lift with a very small drift, so that for a minimum expenditure of power a maximum load carrying capacity is obtained.

Apparatus used to prove existence of lift and drift.—An apparatus used to demonstrate the existence of these forces is shown in Fig. 8. The inclined plane B is fastened to the arm S hinged to the carriage C

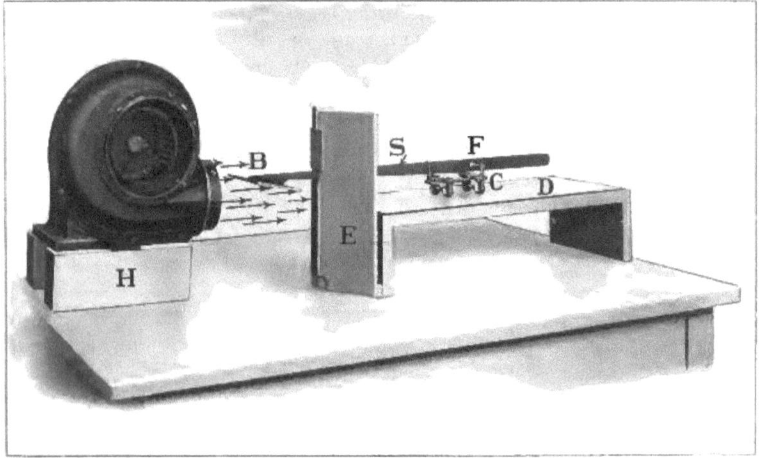

Fig. 8—Apparatus proving existence of both lift and drift

at the point F. The carriage rests on a glass plate D and is shielded from the wind from the blower H by the screen E. It is found that when the blower is started the plane B will lift and the carriage C moves slowly backward carrying the plane with it, thus proving the existence of lift and drift. The screen E is then removed and it is found that the carriage moves away very rapidly thus showing the effect of the added head resistance due to the carriage itself.

Suction on Top of Plane

The flat surface is seldom used for the aerofoils of an airplane. The following illustrations and explanation will help to show the reasons for not using it.

The plane P (Fig. 9) has an opening at O connected to manometer M, while on the under side is a similar opening connected to the man-

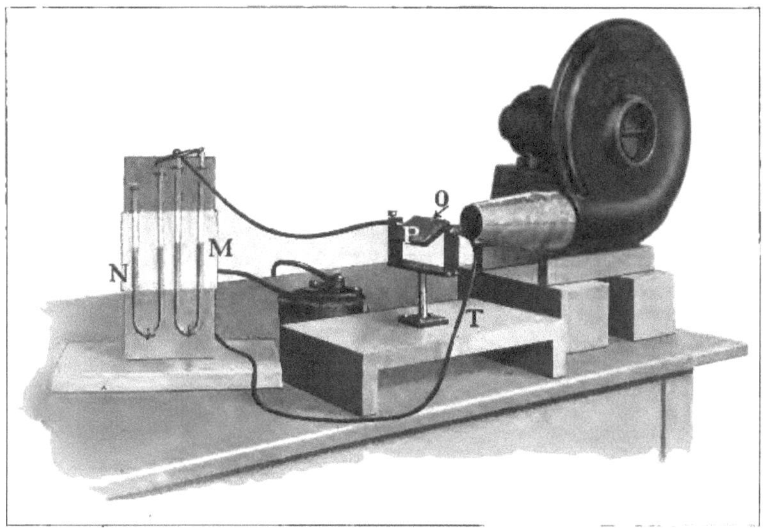

Fig. 9—Device for measuring comparative air pressures on upper and lower surfaces of an inclined plane

ometer N through the rubber tube T. When the blower is started the manometer M shows suction at the point O on the upper side of the plane and N shows pressure on the under side of the plane. In other words, the plane is not only blown up, but it is sucked up as well.

This is very effectively illustrated by a still simpler experiment. Fig. 10 shows the plane AB of heavy cardboard to which is fastened a light strip of paper at the point A and left free at the point C. When the plane is placed in a wind blowing in the direction of the arrows the paper is seen to be drawn up to the position AC′ away from the plane AB.

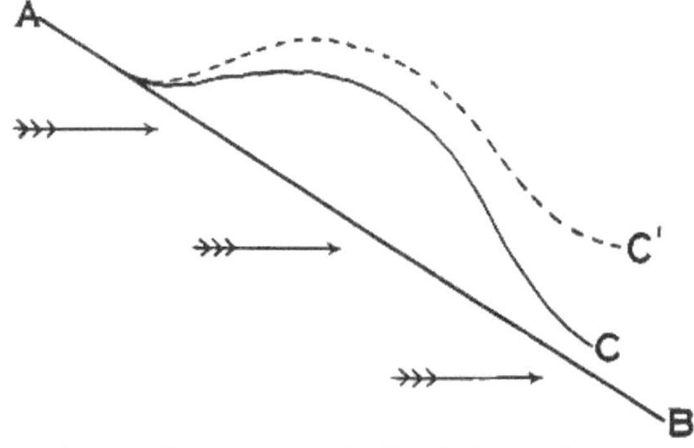

Fig. 10—Showing suction on top of inclined plane when exposed to wind current in direction of arrows

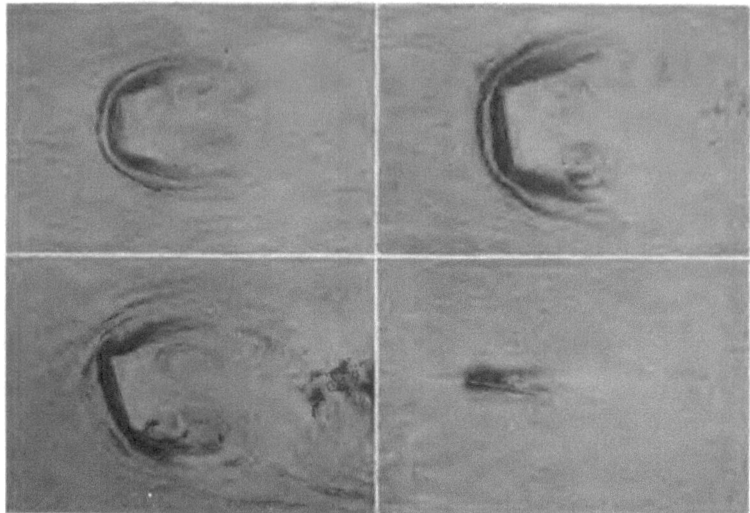

From Loening's *Military Aeroplanes*

Fig. 11—Comparative resistance to advancement of a flat plane at various angles of attack

Fig. 11 represents smoke pictures of a flat plane in four different positions. The lower right hand one shows the plane at a very small angle of attack. The existence of the vacuum at the upper front edge of the plane is very evident.

Experiments at Eiffel Laboratory.—Fig. 12 shows the result of accurate measurements by M. Eiffel of the suction on top of a plane and the

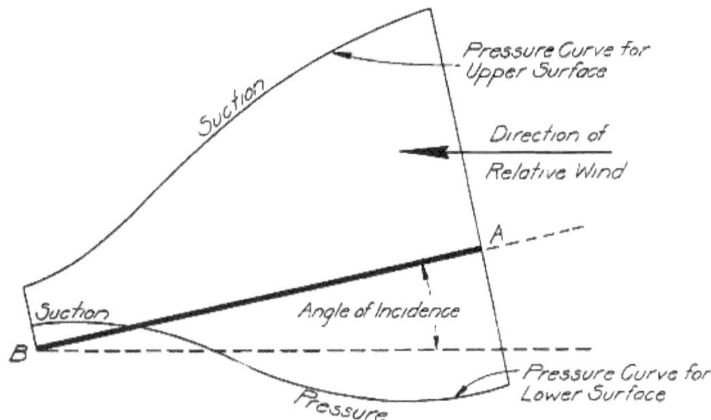

Fig. 12—Pressure diagram of upper and lower surfaces of inclined plane

Fig. 13—In a flat plane, center of pressure C moves toward the leading edge A as the angle of incidence becomes smaller

pressure underneath. Furthermore, Eiffel has shown by recent experiments that when the angle of incidence of a flat plane is low, the value of the suction on the upper surface is considerably more than that of the pressure on the under surface. Thus in this case it is the upper side of the plane which contributes most towards the creation of the lift, a function increasing as the angle grows smaller. This fact shows that the profile of the upper surface of a plane has as much, if not more, importance from the standpoint of the value of lift than that of the under surface.

Center of Pressure

In Fig. 1, it is evident that the wind's force on the plane B could be entirely replaced by a single force acting at the center of the plane. The fact that this point would be the center of the plane is due to the fact that the wind strikes the plane absolutely symmetrically. On an inclined

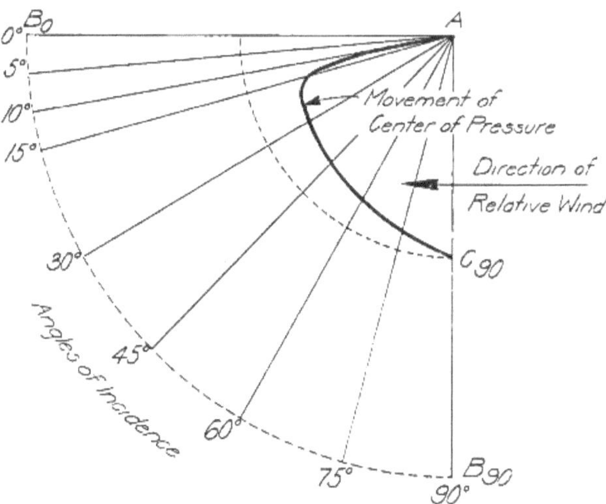

Fig. 14—Location of center of pressure on flat surface for various angles of attack

plane, however, the action of the wind on the front or advancing edge of the plane is different from that on the rear or trailing edge of the plane, hence, we can no longer say that the center of pressure is at the geometrical center of the plane.

The result of the double action of the air-current with pressure below and suction above, both unequally distributed, is that the total reaction on the plane is applied at a point C (Fig. 13) nearer to the leading edge A than to the trailing edge B. This point C is called the center of pressure of the plane. In a flat plane, C moves toward the forward edge as the angle of incidence becomes smaller, until when the angle is zero it reaches the point A.

The curve, Fig. 14, shows the position of the center of pressure on a flat plane for different angles of attack. It will be noticed that from 15 deg. to 0 deg. the center of pressure moves very rapidly towards the front of the plane A. The wind is supposed to be blowing from the right in a direction perpendicular to AB. Airplanes almost never fly with an angle of attack greater than 15 deg. This change in position of the center of pressure very easily can be proven by a well-known and very simple

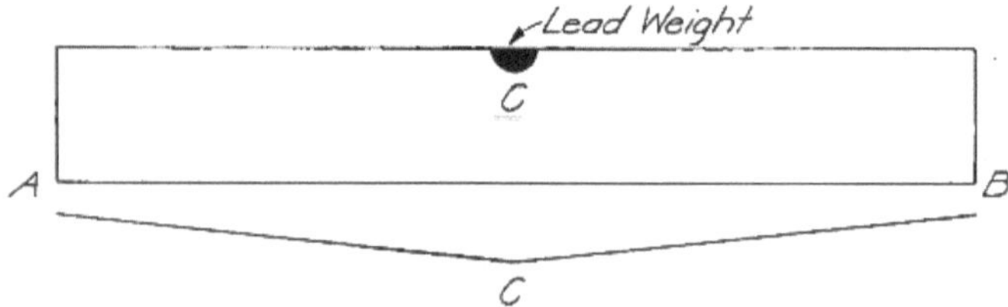

Fig. 15—Center of pressure located close to forward edge of cardboard strip used in simple experiment

experiment. If we take a strip of light cardboard about 8 in. long by 1½ in. wide we know that the center of gravity will pass through the geometrical center. Now if we were to project this through the air in a horizontal position with the long side forward, the center of pressure being at the front end and acting upwards, while the weight at the center of gravity acts downwards, a couple would be produced causing the plane to rotate with the advancing edge going up. This shows that the center of pressure is near the front edge.

We cannot change the center of pressure but we can change the position of the center of gravity by placing a small lead weight on the front edge. Then if the corners at A and B, Fig. 15, are turned slightly upwards while the whole is given a lateral dihedral angle as shown in the lower part of Fig. 15, the plane on being projected in the air is seen to glide almost perfectly. A little practice is necessary in adjusting the weight.

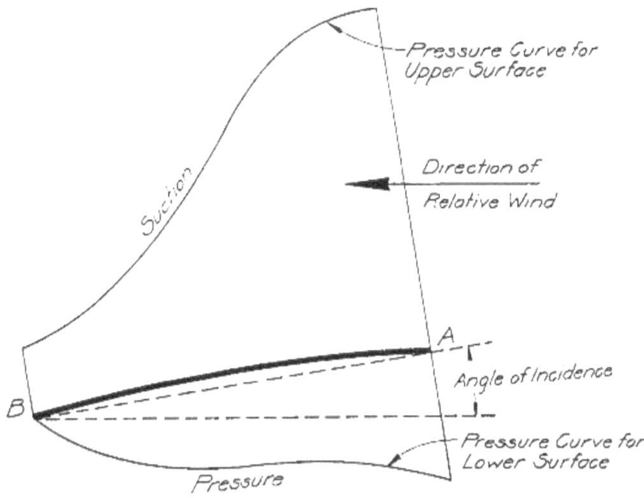

Fig. 16—Pressure diagram for upper and lower faces of curved surface with inclined chord. Compare with Fig. 12

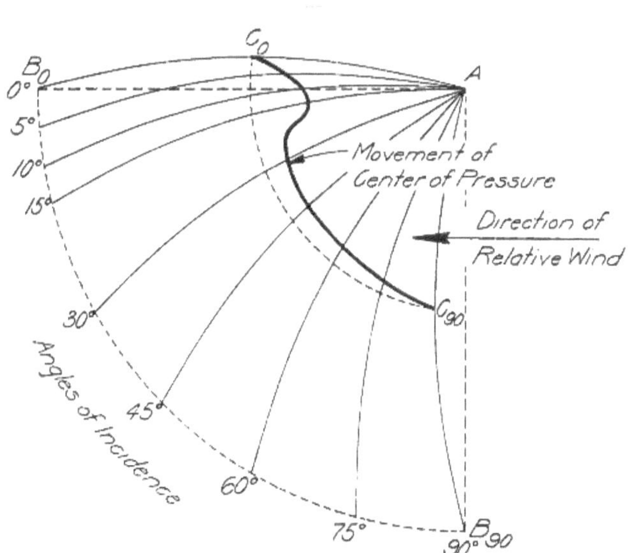

Fig. 17—Location of center of pressure on a curved surface at various angles of attack. Compare with Fig. 14

Figs. 16 and 17 show pressures and the path of the center of pressure for a curved surface. It will be noted first how greatly the suction effect on the top of the plane has been increased, and that from zero to 15 deg. (see Fig. 17) the center of pressure moves in exactly the reverse direction from the way it does in a flat plane. This latter effect has a very important bearing when we come to stability.

Cambered Planes

Fig. 18 is a rough sketch of what one might call a typical wing section. Note the difference in profile between the top and bottom surfaces. The *chord* may be defined as the straight line which is tangent to the under surface of the aerofoil section, front and rear, and the *angle of attack* as the angle between the relative wind and the chord of the aerofoil. We may write the following simple expression for the lift and the drift:

$$\text{The Lift (L)} = k_L S V^2$$
$$\text{The Drift (D)} = L/r.$$

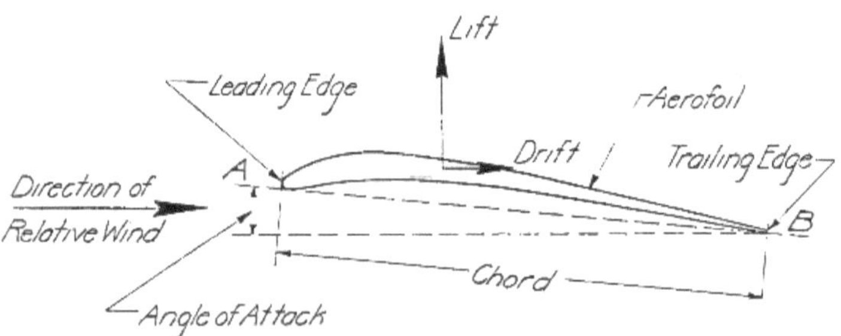

Fig. 18—Sketch of a typical wing section with aeronautical terms indicated

The coefficient k_L depends upon the shape of the aerofoil and the angle of attack and must be determined experimentally. The quantity r, also determined experimentally, is called the lift-drift ratio and measures the efficiency of the aerofoil.

Fig. 19, gives two curves for an aerofoil of the section shown. The first curve gives the values of the quantity k_L for different angles of attack, while the second curve gives the values of the lift-drift ratio.

For example, suppose that an airplane with aerofoils of the type shown, lifting surface 60 sq. ft., is flying at an angle of attack of 11 deg., and with a velocity of 70 m.p.h. What will be the lift and the drift?

From the chart, Fig. 19, we find that for this type of plane and angle of attack $k_L = 0.0028$ and $r = 11$, hence,

$$L = k_L S V^2 = 0.0028 \times 60 \times (70)^2 = 823 \text{ lbs.}$$

$$D = \frac{L}{r} = \frac{823}{11} = 75 \text{ lbs.}$$

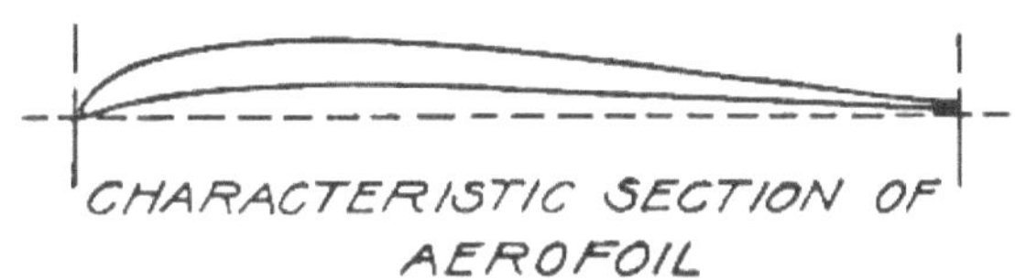

CHARACTERISTIC SECTION OF
AEROFOIL

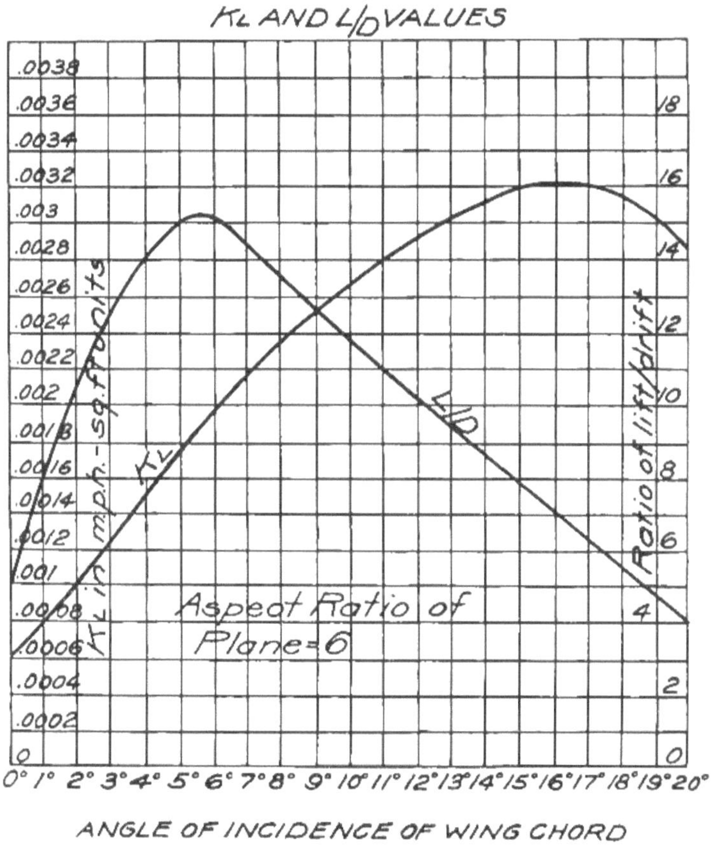
KL AND L/D VALUES
.0038
.0036
.0034
.0032
.003
.0028
.0026
.0024
.0022
.002
.0018
.0016
.0014
.0012
.001
.0008
.0006
.0004
.0002
0
18
16
14
12
10
8
6
4
2
0
KL in m.p.h.-sq.ft units
KL
L/D
Ratio of lift/drift
Aspect Ratio of
Plane=6
0° 1° 2° 3° 4° 5° 6° 7° 8° 9° 10° 11° 12° 13° 14° 15° 16° 17° 18° 19° 20°
ANGLE OF INCIDENCE OF WING CHORD

Fig. 19—Curves showing values of k_L and Lift/Drift ratio for a typical wing section

If now we change the angle of attack to 3½ deg., keeping the surface and velocity the same, we find from the chart that $k_L = 0.0014$ and $r = 13.5$, hence,

$$L = k_L SV^2 = 0.0014 \times 60 \times (70)^2 = 412 \text{ lbs.}$$

$$D = \frac{L}{r} = \frac{412}{13.5} = 30 \text{ lbs.}$$

Horizontal Flight

For horizontal flight the lift produced by the machine's velocity must at all times exactly equal its weight. For if the lift were less than the weight of the plane would fall, while if the lift were greater than the weight the machine would begin to climb. We therefore can replace the lift by the weight W. Then we would have for horizontal flight:

$$\text{Weight (W)} = k_L SV^2$$
$$\text{and the drift (D)} = W/r.$$

For example, a given airplane weighs (with load) 1800 lbs. Its aerofoils are of the type illustrated and the lifting surface is 120 sq. ft. What will be its velocity for horizontal flight at an angle of attack of 12 deg.?

From the chart, Fig. 19, we find that for this type of plane and angle of attack, $k_L = 0.0029$, whence,

$$L = W = k_L SV^2 \text{ or } 1800 = 0.0029 \times 120 \times V^2$$

$$\text{transposing, } V^2 = \frac{1800}{.0029 \times 120} = 5172$$

$$\text{hence, } V = \sqrt{5172} = 72 \text{ m.p.h.}$$

If now we reduce the angle of attack to 5 deg., the chart, Fig. 19, shows that k_L becomes 0.00175, whence,

$$1800 = 0.00175 \times 120 \times V^2$$

$$\text{transposing, } V^2 = \frac{1800}{0.00175 \times 120} = 8572$$

$$\text{hence, } V = \sqrt{8572} \text{ or } 92+ \text{ m.p.h.}$$

The above example illustrates this important principle that, since a machine in horizontal flight, except for a slight loss due to consumption of gasoline, maintains a constant weight and a constant surface and since k_L for a given plane depends solely upon the angle of attack, the velocity for horizontal flight is completely determined when we know the angle of attack. Now since the pilot can control the angle of attack by means of his elevators he can control the velocity for horizontal flight.

Fig. 20 shows four different positions of the plane corresponding to four different angles of attack. In each case the machine is flying horizontally, though at first sight one might think that in position 4 the machine was climbing.

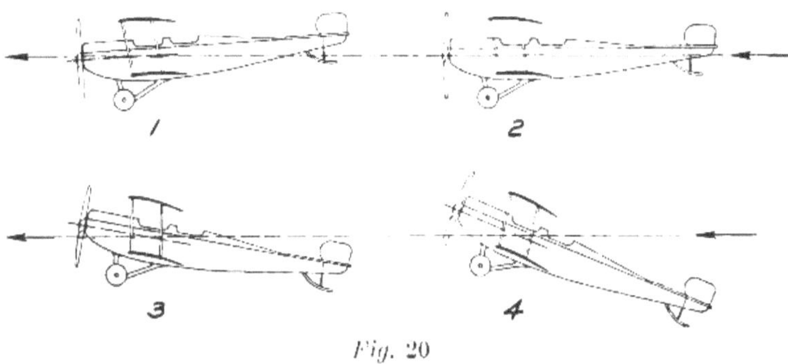

Fig. 20

FOUR POSITIONS FOR FLIGHT

(1) *Minimum angle.*—This is the smallest angle at which horizontal flight can be maintained for a given power, area of surface, and total weight. The minimum angle gives the maximum horizontal flight velocity at low altitude. Note that the propeller axis is inclined slightly downwards when flying at this angle.

(2) *Optimum angle.*—This is the angle at which the lift-drift ratio is highest. In modern airplanes the propeller axis is generally horizontal at the optimum angle, as shown at (2) in the above figure. Note that in the position shown the velocity of the airplane will be less than when flying at the minimum angle. The effective area of wings and angle of incidence for the optimum angle are such as to secure a slight climbing tendency at low altitude.

(3) *Best climbing angle.*—This angle is a compromise between the optimum and maximum angles. Modern airplanes are designed with a compromise between climb and horizontal velocity. At this angle the difference between the power developed and the power required is a maximum, hence the best climb is obtained at this angle. See Fig. 22.

(4) *Maximum angle.*—This is the greatest angle at which horizontal flight can be maintained for a given power, area of surface and total weight. If the angle is increased over this maximum, the lift diminishes and the machine falls.

It would seem at first that we have entirely neglected the engine, especially as there is a general impression that the velocity of a machine depends upon the power of the engine, while as a matter of fact the form of wing sections together with the plane's dimensions are equally, if not more, important. In the preceding discussion we have simply assumed that the engine had the necessary power to maintain the plane at such a velocity as was determined by that angle of attack at which the pilot drives the machine.

Engine Power

The power of any engine is measured by the velocity at which it can move a body against a given resistance, and its unit, the horsepower, may be defined as the power required to lift one pound 33,000 ft. in one minute or 375 miles in one hour, or the power required to lift 375 pounds one mile in one hour.

We must therefore multiply the total resistance offered to the airplane, which consists of the drift plus the parasite resistance multiplied by the velocity of the machine, and divide the result by 375 to get the horsepower required. Or, written as a formula:

$$\frac{(\text{Drift} + \text{Parasite Resistance}) \times \text{Velocity}}{375} = \text{Horsepower.}$$

From the above expression for horsepower, it will be noted that since the drift for a given machine depends solely upon the angle of attack, and the parasite resistance depends upon the square of the velocity, which

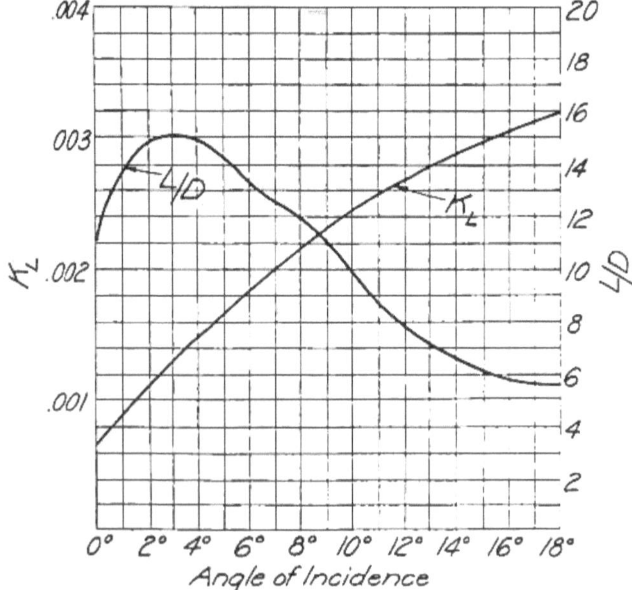

Fig. 21—Value of K_L and Lift Drift ratio for a given machine

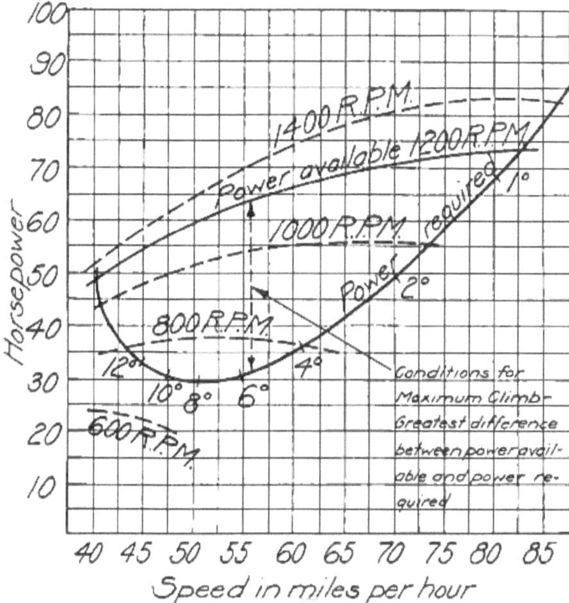

Fig. 22—Showing power required at different angles, also power delivered

in turn depends upon the angle of attack, we may state that for a given machine with its load, the horsepower is completely determined when we know the angle of attack at which the machine flies.

Fig. 21 corresponds for the entire machine to Fig. 19 for the aerofoil itself and gives the value of k_L for a given machine, as well as the lift-drift ratio.

Fig. 22 gives in the heavy curve the power required to drive the machine at the angles of attack marked on the curve, which correspond to the speed in miles per hour given at the bottom. The other set of curves, four of them dashed and one a light line, give the power delivered to the machine by the engine through the propeller. The latter would be straight horizontal lines were it not for the fact that the efficiency of the propeller varies with the velocity of the airplane. The ordinates as shown on the left side of the diagram correspond to horsepower.

Let us consider the case where the engine is making 1200 r.p.m. It will be seen that if the pilot changes his elevators so as to fly with an angle of attack of a little less than 1 deg., or of a velocity of about 82.5 m.p.h., he will be using every particle of power that his engine can deliver at that speed. Any slight decrease in the angle of attack will cause him to go down probably in a nose dive. As he increases the angle of attack we come to a point where the distance between the two curves, power delivered and power required, is the greatest. Here we will have

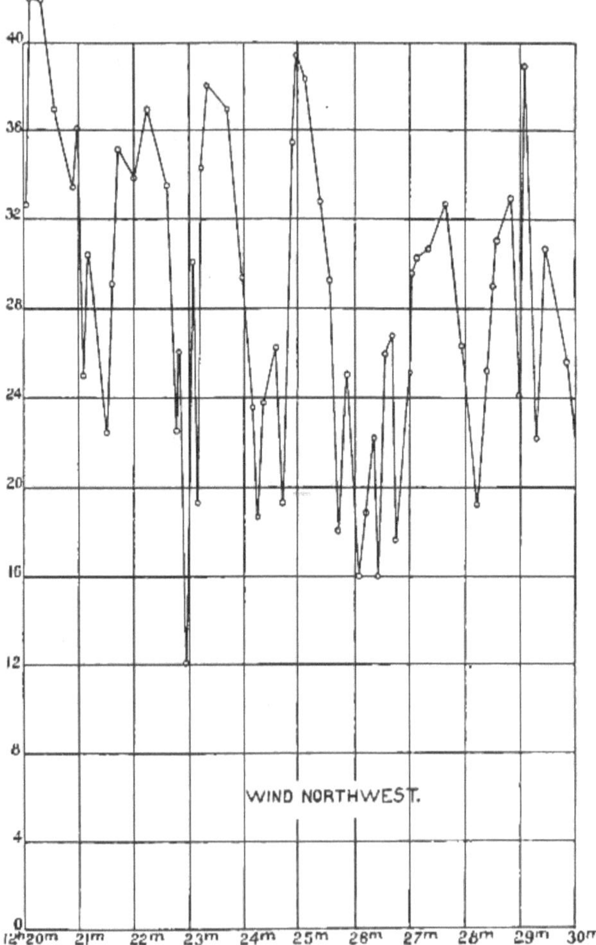

Fig. 23—Showing rapid changes in wind velocity in short spaces of time

the greatest excess of power over that used for horizontal flight, all of which can be used in climbing. Hence that point will be the position for maximum rate of climb. It is indicated by the vertical dash line marked *maximum climb* at an angle of attack of a little less than 6 deg.

or a velocity of a little over 55 m.p.h. Increasing his angle of attack still further, or at about 8 deg., which is the lowest point on the curve, where the horsepower required for horizontal flight is only 30, we get a point of most economical flight. Then, as we decrease the angle of attack, the power required rises rapidly until at 40 m.p.h. the two curves cross again and any increase in the angle of attack would cause the machine to stall in the sense of going down, which might take the form of either a nose dive or tail slip. It is well to compare this with Fig. 20.

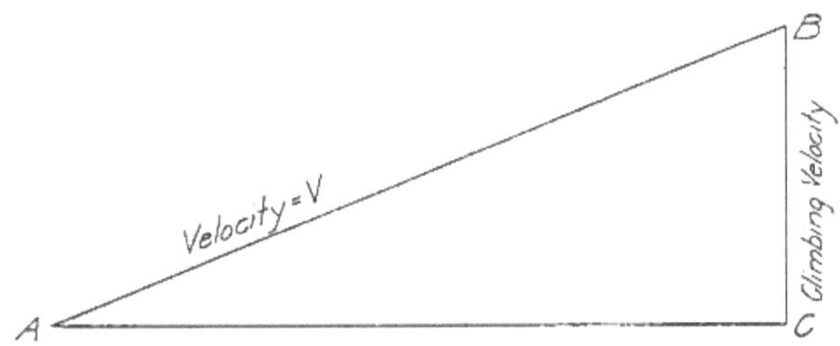

Fig. 24—Calculation of power required to climb

It is also interesting to compare this with Fig. 23, taken from Langley's *The Stored Energy of the Wind*, and which illustrates the rapid changes in the velocity of the wind occurring in short intervals of time. The vertical lines represent spaces of one minute and the horizontal lines wind speeds differing by 4 m.p.h. It will be noticed that between 32 and 24 min. the wind fell from about 37 m.p.h. to 12 m.p.h. and rose again to 38 m.p.h. On account of the momentum of the airplane it would be practically impossible for its actual velocity to change with anything like that rapidity, and as the lift depends upon the square of the velocity it is evident that the pilot would experience a series of "bumps" when the velocity increased, and momentary drops when the velocity decreased. The feeling has been likened to a motor boat driving rapidly through a choppy sea.

Power to Climb

Suppose the center of gravity of a machine be moving in the direction AB, Fig. 24, with a velocity of V miles per hour. The horsepower will then be the sum of two components, viz., that necessary to overcome the wind resistance, as already given for horizontal flight and that necessary to lift the machine through the distance CB in the time required for the machine to travel from A to B. Now if AB be taken to represent the distance the machine travels in an hour, BC would then represent the velocity of climb. The power consumed in climbing is equal to the product of the weight of the machine in pounds by the velocity of climb in miles per hour divided by 375. Let us call AB/BC the climbing ratio R which gives us BC = AB/R = V/R. We will have then the power expended in the climb alone equal to WV R, and the total horsepower becomes:

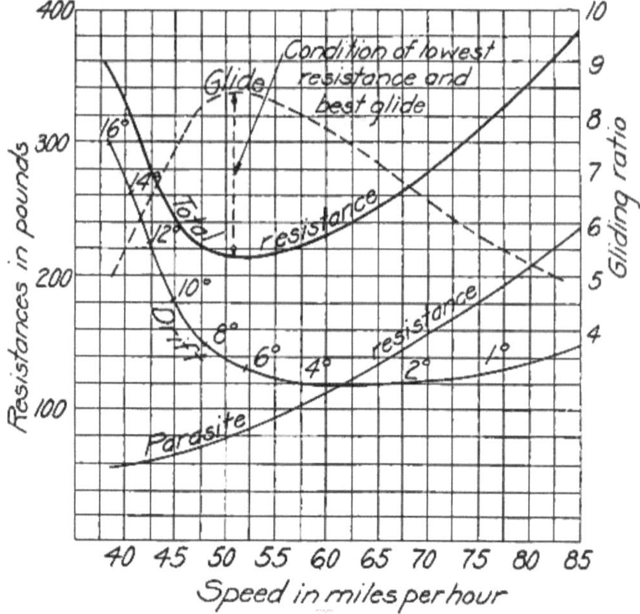

Fig. 25—Showing how drift, parasite resistance and gliding motion depend upon angle of attack

$$\text{Horsepower} = \frac{(\text{drift} + \text{parasite resistance})V}{375} + \frac{WV}{375\,R}$$

The case of special interest is where the horsepower becomes zero. This is the condition when the engine is shut off on a glide.
When,

$$\frac{(\text{drift} + \text{parasite resistance})V}{375} + \frac{WV}{375\,R} = 0, \text{ this reduces to}$$

$$R = -\frac{W}{(\text{drift} + \text{parasite resistance})}$$

It should be noted that the value of R is negative, due to the fact that the machine is gliding toward the earth. Now since both drift and parasite resistance depend upon the angle of attack, the gliding velocity and slope depend upon the angle of attack, and are under the control of the pilot. This is illustrated in Fig. 25.

Stability

One of the most important considerations in an airplane is stability, which is generally considered under three headings, viz., longitudinal, lateral and directional.

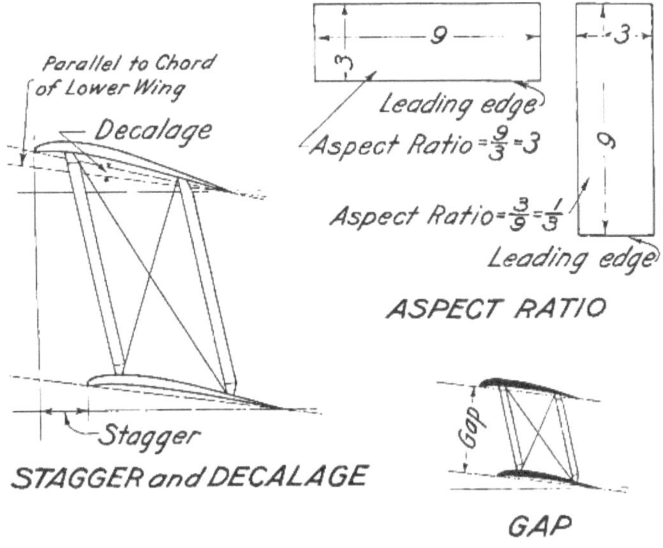

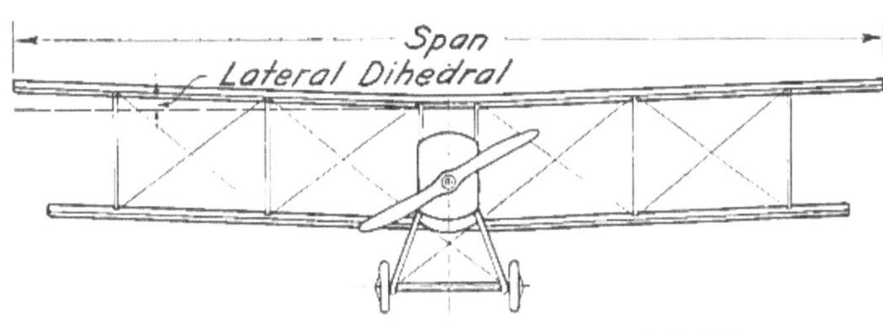

LATERAL DIHEDRAL and SPAN

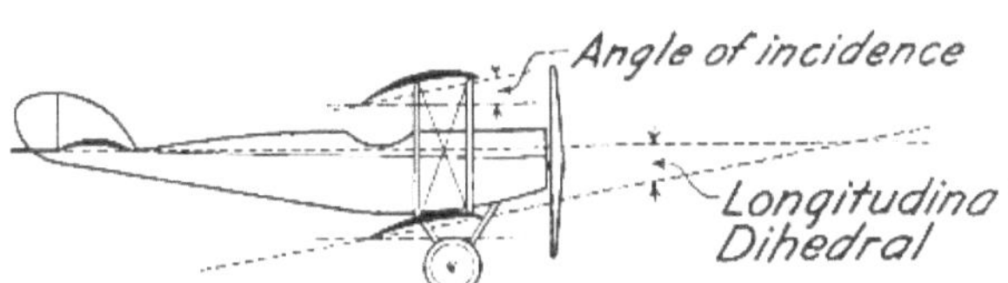

LONGITUDINAL DIHEDRAL

Fig. 26—Illustrating meaning of some aeronautical terms

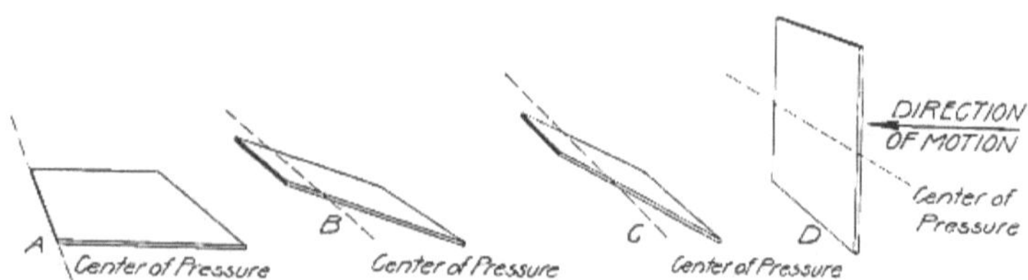

Fig. 27—The center of pressure of a flat plane moves forward as the angle of incidence is decreased

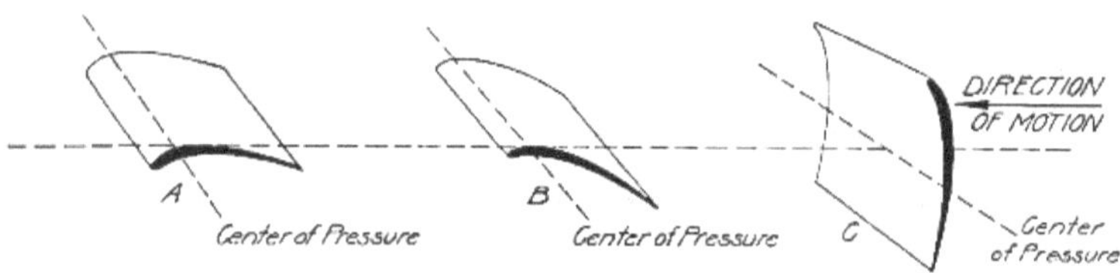

Fig. 28—The center of pressure of a curved surface moves forward with decreasing angles of incidence up to about 12 deg. Below this angle it reverses and moves toward the center again

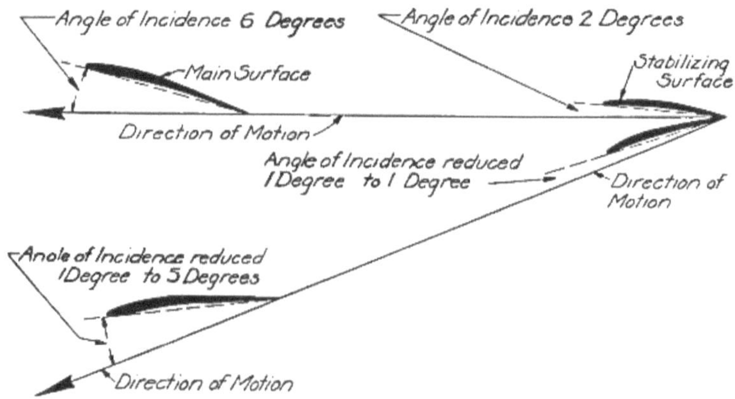

Fig. 29—Illustrating how the rear surface has its angle of incidence reduced in greater proportion than does the front surface when the combination is tipped downward

Longitudinal stability.—This stability is needed to keep the airplane from pitching nose downward or tipping backward, nose up and tail down, whenever a gust or eddy is encountered.

Flat surfaces are longitudinally stable because, as shown in Fig. 14, the center of pressure moves toward the leading edge as the angle of incidence is decreased. Fig. 27 shows four positions of a flat surface moving from right to left. Moving horizontally as in position A the center of pressure is at the leading edge, and when in the vertical position D the center of pressure coincides with the transverse center line of the surface. However, suppose the surface to be moving as at C and a sudden gust of wind tips it into position B with a lesser angle of incidence. Then the center of pressure moves forward, introducing a greater moment and tending to force the plane back into its original position C. On the other hand, if the surface assumes too great an angle, the center of pressure moves back and the rear is forced up, causing the surface again to

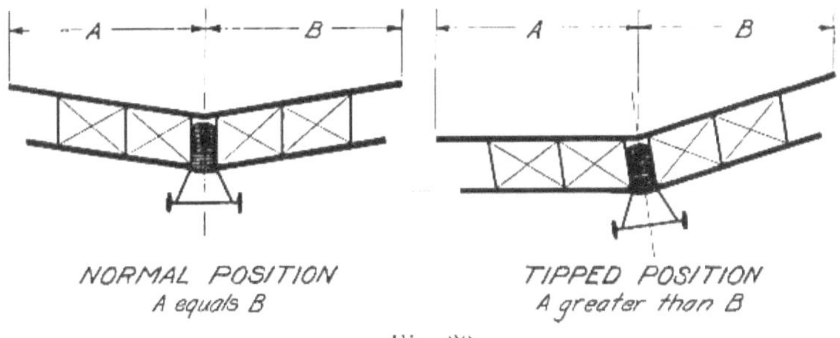

Fig. 30

resume its original position C. Thus, if it were not for the fact that the flat surface has a very poor ratio of lift to drift, it could be used in airplanes to advantage, due to this inherent longitudinal stability.

Next consider Fig. 28, giving three positions of a cambered surface, which has a much greater lifting efficiency than a flat surface. It is also supposed to be moving from right to left. In position C the center of pressure coincides with the transverse center line. Supposing this surface to be moving in attitude B with the center of pressure at approximately the position indicated. If it is suddenly tipped into position A, it will be seen that the front part has a negative angle of incidence, which results in a downward pressure on this portion. The center of pressure of the surface being the resultant of all forces acting, it is obviously affected by this action at the front, and moves backwards. If the surface is tipped still further, the backward movement of the center of pressure is increased and therefore there is still less tendency to push the front up, when such a tendency would be most desirable. On the other hand if the angle of incidence becomes suddenly greater than the normal position B, the pressure on the front edge decreases and the resultant center of pressure moves forward, thus tending to push the front up and give the surface a still greater angle of incidence.

Therefore, it is necessary to have some way of compensating for this instability of cambered surfaces, and this is done by the use of an auxiliary stabilizing surface some distance back from the main surface and set at a lesser angle of incidence than the main surface. Such a stabilizer is a necessary feature of all modern airplanes. Fig. 29 shows two such surfaces in tandem, thus forming an elementary airplane. Consider the airplane to be traveling horizontally with the angle of incidence of the

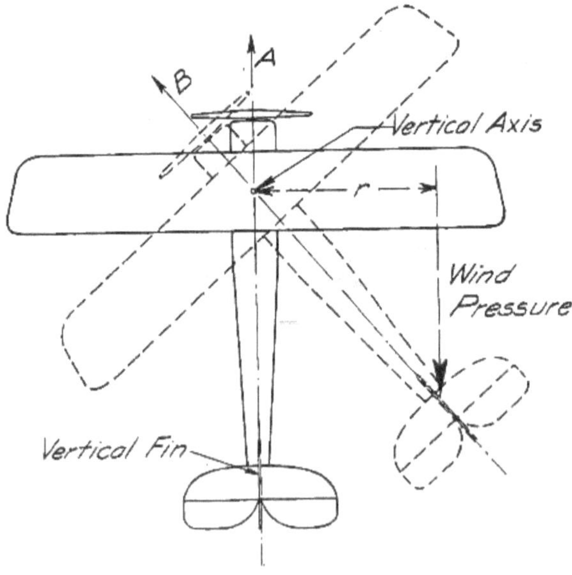

Fig. 30A—Diagram to show action of vertical fin in preserving directional stability

main surfaces 6 deg. and the rear one-third of this, or 2 deg. Now supposing a sudden gust pitches the plane into some such position as shown in the lower part of the diagram. The angle of incidence of both surfaces is now reduced say 1 deg., the main surface being at a 5 deg. angle and the rear surface at 1 deg. In other words, the main surface has lost only about 17 percent of its angle of incidence, whereas the stabilizer has lost 50 percent. Consequently the stabilizer has lost more of its lift than the main surface, and it therefore must fall relative to the position of the main surface, bringing the combination back into normal position again. On the other hand, if the front of the plane is suddenly forced up, the stabilizing surface receives a relatively greater increase in angle of incidence than the main surface, hence relatively greater increase in lift, causing the back end of the plane to be brought up until the combination again is normal.

Lateral stability.—This stability is necessary to prevent the machine from rolling about its horizontal axis. It is difficult to secure, but is often promoted by having a slight lateral dihedral angle between the upper wing surfaces, as shown in Fig. 30. Should the airplane suddenly be tipped to one side, in the position shown to the right of the diagram, the planes on the down side become more nearly horizontal, whereas, those on the other side assume an angle still greater than they had when flying normally. Thus, the effective projected lifting surface on the side A is increased and that on side B is decreased, bringing the plane back to its normal lateral position. Other features are introduced to aid lateral stability, such as "wash in" on the left side to give this side slightly more lifting ability to compensate for the torque of the propeller.

Directional stability.—Such stability aids in keeping the plane on its course. In order to prevent yawing with every gust of wind, the vertical tail fins present on nearly all modern planes are used. Referring to Fig. 30 A, suppose a sudden gust of wind to deflect the airplane from its normal course A so that the nose points off the course to the pilot's left, as indicated by the dotted lines in position B. This swings the tail around to the right so that the right side of the vertical fin presents a flat surface to the wind pressure resulting from the tendency of the machine still to move forward in the direction A, due to its inertia, even though it is temporarily pointing in direction B. A moment with arm r is thus set up, which tends to swing the plane back on its vertical axis until the fin is again parallel to the direction of the relative wind. The action is similar to that of a wind vane, the vertical fin of which always keeps it pointing in the direction of the wind.

Chapter 2

TYPES OF MACHINES

General divisions—Dirigibles and balloons—Heavier-than-air craft—Training machines, primary and secondary—Pursuit planes—Reconnoissance machines—Bombing and raiding machines, day and night.

ALL aircraft are divided into two general classes: heavier than air and lighter than air. In the lighter-than-air class (which consists of aircraft supported in the air by the buoyancy of a gas, lighter than the air it displaces, contained in a gas bag of some convenient form), a further sub-division may be made into dirigibles, or craft equipped with a power plant, propeller and vertical and horizontal rudders, so that they can be operated and steered at the will of the pilot, and balloons, or craft not fitted with any means of propulsion or steering, and which therefore drift with the wind unless held captive by means of a cable anchoring them firmly to a point on the ground.

Dirigibles are further divided into three types, the rigid, the semi-rigid and the non-rigid, all three of which are being used to a considerable extent in the present war. The Zeppelin, which has proven such a costly failure for Germany, is the outstanding example of the rigid construction. The Blimp as used so successfully by the British in coast-guard observation and anti-submarine work, and as being made in this country, is a good example of the non-rigid type. There are several examples of the semi-rigid construction, such as the German Parseval and others.

In the balloon, or non-dirigible class, the captive observation balloon and particularly the kite type balloon are being employed to a considerable extent, even though they may fall an easy prey to enemy airplanes unless strongly protected. With improved methods of protection, by means of anti-aircraft barrage, the employment of winches which rapidly pull the captive balloon to cover, and a thorough protective patrol by scouting airplanes, the balloon as a means of reconnoissance and for long distance artillery observation has come into increased favor. Important improvements are also being attempted by using non-inflammable gas in the balloon, or by fitting a protective housing consisting of an outer bag containing an inert gas such as nitrogen.

In any case the observer in a captive balloon is always equipped with a parachute ready for immediate use in case the balloon should be destroyed by the enemy by any means before it could be pulled down.

Aircraft Heavier Than Air

There are four types of heavier-than-air machines: airplanes, kites, helicopters and ornithopters. Only one of these, the airplane, for practical reasons is worth considering at this time. This type, being of greatest importance will be studied both from the standpoint of its uses and its constructional features.

Airplanes for military use are divided into five general types: training, pursuit, combat, reconnoissance and bombing or raiding.

Training machines are either primary or secondary. The former is used in elementary training and is generally of the dual control type, so as to allow the instructor to control the machine until the pupil becomes accustomed to the "feel" of it, under the guidance of an experienced pilot, then take over the control gradually, and to allow the instructor to correct mistakes of the pupil before they can have serious consequences. Machines called "rollers" or "penguins," having curtailed wings to prevent them actually rising off the ground, are also used in primary instruction. Secondary school machines are generally similar to those used for actual fighting work; this being particularly true of scout or pursuit machines for "stunt" flying. Training machines should be easily handled, should possess marked inherent stability and should have a fairly slow get-away speed. A familiar machine used in this country has the following characteristics:

Tractor biplane
Two seater
Horsepower—80 to 120.
Radius of flight—200 mi.
Rate of climb—300 ft. per min.
Minimum flying speed—45 m.p.h.
Maximum speed—75 m.p.h.

In France and England, quite a number of Farman and B. E. pusher machines are used for training, and for advanced work the Nieuport Scout, the Bleriot monoplane and similar machines are used, especially in French schools.

Pursuit planes.—This class comprises the fastest and most easily handled machines that it is possible to produce. Their offense depends on speed and their defense on ability to maneuver. Due to the great strains imposed in "stunt" flying, the monoplane has lost favor on account of structural weakness. The Morane monoplane which is still in use is, however, an exception. The Nieuport one-and-one-half plane,—probably the most successful pursuit machine—the Spad, the English Bristol and the Sopwith scouts are all popular with Allied aviators. The prime requisites for scouts are speed, ability to climb and power to maneuver. Scouts may be either single or two-seaters. They always carry one gun either fixed and firing through the propeller or on the upper plane. Other gun mountings may be used, however, especially when an extra passenger is carried. The principal characteristics of scouting machines are:

Tractor biplane
Horsepower—above 150
One or two-seater
Radius of flight—300 mi.
Rate of climb—over 800 ft. per min.
Minimum flying speed—50 m.p.h.
Maximum speed—150 m.p.h.
Ceiling—20,000 ft.

Combat machines.—Airplanes of the combat type are used extensively for strictly fighting purposes, and are essentially the same as the reconnaissance type machines except that they are stripped of wireless equipment, photographic apparatus and other accessories not essential for fighting purposes. The combat machine is a two-seater and carries four guns, two in the observer's cockpit, and movable on a circular track surrounding the cockpit, and the other two fixed and synchronized to fire between the propeller blades and operated by the pilot. These machines usually have a ceiling of between 20,000 and 23,000 ft. and carry oxygen tanks for the passengers. They are of the tractor biplane type, and considering their weight and fighting ability, have remarkable maneuverability. The principal features of combat planes are:

Tractor biplane
Horsepower—250 or more
Two passengers
Radius of flight—300 miles or more
Rate of flying—10,000 ft. in 10 min.
Minimum speed—50 m.p.h.
Maximum speed—150 m.p.h. or over
Ceiling—20,000 to 23,000 ft.

The reconnoissance machine.—This type, usually carrying an observer, wireless, photographic apparatus and sometimes a number of light bombs is usually armed with one or two machine guns. The purpose of this type is to do various forms of scout and observation work both above and behind the lines, and also contact patrol work. These machines fly at altitudes of from 2,000 to 6,000 feet and usually rely on the pursuit machines for protection. This class of machines is one which comprises a large assortment of constructions. They are generally biplanes, pusher or tractor, and quite often with single, double or triple motors. The armament consists generally of two machine guns, one mounted fixed and firing ahead, the other movable and operated by the observer. The general qualities of these planes are as follows:

Biplanes—tractor, pusher or combination
Two passenger or more
Horsepower—200 or over
Radius of flight—300 mi. or over
Rate of climb—200 ft. per min. or over
Minimum flying speed—50 m.p.h.
Maximum speed—110 m.p.h. or over

Bombing or raiding.—These machines are large, slow, weight-carrying planes. In order to get the latter quality, a biplane, triplane or even a multiplane construction is necessary since there is a limit to the span of wings. Parasite resistance is high and horsepower must necessarily be large. This form of machine is rather new and has been developed during the recent war, because of its wonderful possibilities, and it is only reasonable to suppose that very marked improvements will come in the future. The larger Handley-Page (British) bombers, and the Italian Caproni triplanes are an indication of what developments are being made in raiding machines. If the Allies are successful in clearing the air of German planes, any destruction or offensive operations must be accomplished by the bomber. The extent of damage which might be inflicted in this way is limited only by the number of machines and the amount of bombs dropped. These planes rely on the accompanying pursuit planes for protection. Raiding or bombing expeditions are always carried out in formation and the number taking part is unlimited. The characteristics likewise are without limit.

The principal features of the large bombing planes are as follows:

Biplane, Triplane
Horsepower—no upper limit. (As many as five engines are being used.)
Number of passengers—from two up
Range of action—over 300 mi.
Weight carried—above 1000 lbs.
Rate of climb—250 ft. per min. and over
Minimum flying speed—45 m.p.h.
Maximum speed—up to 85 m.p.h.
Ceiling—10,000 ft.

A further classification of day and night bombers is made. Night work is dependent on suitable lighting signaling arrangements, proper landing signals and the ability to reckon position in the dark. The Germans have given considerable attention to this branch and it is also being practiced by the Allies.

Chapter 3

SHIPPING, UNLOADING AND ASSEMBLING

Shipping instructions—Marking boxes—Methods of shipping—Railroad cars used—Unloading—Method of loading on truck—Tools required—Unloading from truck—Unloading uncrated machines—Opening boxes—Assembling—Fuselage and landing gear—Center panel and wings.

SHIPPING *instructions.*—Boxes in which airplanes or parts thereof are shipped should be marked with the following:

Destination, or name and address of consignee in full.

Sender's name.

Weight of box (gross, net and tare).

Cubic contents (or length, width and height).

Box and shipment number.

Hoisting center.

"This side up."

Methods of shipping machines. — Machines are shipped either by loading in a railroad car without crating, or by crating in two boxes. In the latter case the wings, center-section panel, tail surfaces, landing gear and propeller are removed from the fuselage, and the fuselage, landing gear, propeller and radiator are packed securely in the fuselage box. The other parts are packed in the panel box. All aerofoil sections are stood on their entering edges and securely padded to protect their coverings. Struts are stood on end.

If the machine is not to be crated only the following parts are removed—wings, center section panel, tail surfaces and propeller. The fuselage is loaded into the railroad car and allowed to rest on the landing gear. The latter should be blocked up, however, to take the load off the tires of the landing gear wheels and off the shock absorbers. The fuselage must of course be securely fastened in the car to prevent movement in any direction. The wings and other separate parts are crated against the sides of the car. The wings are secured with their entering wedges down and carefully padded to prevent damage.

Railroad cars used for transportation.—If possible open end or automobile cars are used for transportation of airplanes. Sometimes with crated machines gondola cars are used, and with uncrated machines.

ordinary box cars having no end doors. In the latter case, however, it is necessary that the side doors of the railroad car be as wide as possible, to allow working the fuselage in and out without damage.

For transporting machines (either crated or uncrated) from the railroad, a flat top truck is used. If the truck is short it will be necessary to use a trailer to support the overhang of the boxes.

Unloading

Method of loading on truck.—Before unloading a machine, everything in the railroad car should be inspected for loss or damage. If everything is O. K. proceed with the unloading, but if any loss or damage is discovered report fully at once to the receiving officer and await his instructions before doing anything further.

The tools required for removal of airplane boxes from the railroad car are: 1 axe or hatchet, 2 crow bars, 6 or 8 rollers and 100 ft. of 1 in. rope.

The cleats holding the boxes to the car floor are first removed with the axe and crow bars, and the panel box removed from the car. If the fuselage box is not marked to show which is the front end it should be lifted slightly, if possible, first at one end and then at the other, to determine which is the engine end. This end, being the heavier, should come out first if possible.

The truck is backed up to the door of the car, rollers are placed under the fuselage box and it is then rolled out onto the truck. The rope is now used to fasten the box to the truck. After this is done the truck is moved forward slowly and the box is thus pulled out of the car. If a trailer is to be used it should be placed under the box before the latter is taken all the way out of the car.

When taking the fuselage out tail end first, the same methods are used, except that the light end is blocked up when removed from the car and a truck is put under the heavy end.

When moving along roads care should be taken to go slowly over rough places, tracks and bad crossings. It is also a good policy to have a man on each side of the box to watch the lashings and see that nothing comes loose.

Panel Box

The wing box (or panel box) is removed from the car in the same manner as the fuselage box.

Unloading boxes from truck.—For this work 2 planks about 2 in. x 12 in. x 12 ft. long should be used. These should be fastened to the end of the truck with one end resting on the ground, so that they will act as skids. The tail end of the fuselage box is depressed until it rests on the ground, then by moving the truck forward carefully the box will slide down the planks onto the ground.

Unloading uncrated machines.—In this case all of the smaller parts should be removed first. Then the cleats and ropes are removed which hold the machine in the car. Two long planks are placed from the door of the car down to the ground and are used to roll the machine out of the car.

Opening boxes.—A screw driver and bit brace should be used to remove the screws in the top, sides and ends of the box. The top is removed first, then one side. All smaller parts of the machine should be taken out, after which the remaining side of the box is removed, and lastly the ends.

Assembling a machine.—The landing gear should be put on first. To do this the fuselage must be raised by one of two methods. The first is by chain falls or block and tackle. The rope sling should be passed under the engine sill just to the rear of the nose plate. The tail of the machine is allowed to rest on the tail skid while the nose is raised. The second method is by shims and blocking. This latter method is the most common because chain falls are not always available. Enough blocks should be secured to raise the fuselage high enough to slip the landing gear underneath. The tail is first raised by 2 men and blocks are placed under Station 5 or the rear wing section strut. The blocking must be directly below the strut and must have padding upon it. Then the tail is depressed and another block is put under the forward wing strut. This operation is then repeated until the fuselage is high enough for the landing gear when the machine is blocked under nose and tail and the other blocks are removed. Three or four men are all that should be required for this second method.

Assembling Wings

After the landing gear is assembled the center section panel should be attached and approximately lined up. Then the wings are assembled. There are two methods for doing this; one is to put on the top planes, place supports under the outer edges, then put in struts and lower planes and connect up the wires. The other method is to assemble the wings completely while on the ground. Wings are stood on their entering edge, struts are put in and wires tightened up to hold the wing sections together. Then the wings are attached to fuselage by turning them over and attaching the top wing first, then the lower wing. One side of the machine must be supported until the opposite set of wings is attached. After wings are all attached, then the tail surfaces should be assembled to the body. The horizontal stabilizer should go on first, then the vertical fin, rudder and elevators in the order named. On some machines the elevators will have to be put on before the rudder. After everything is assembled the machine is put in alignment.

Chapter 4

RIGGING

Fuselage—Construction—Longerons—Struts—Fuselage covering—Monocoque—Landing gear—Struts—Bridge—Axle box or saddle—Axle and casing—Wheels—Tail skid—Shock absorber—Wing skids—Pontoons on seaplanes—Flying boat hull—Wing construction—Front and rear spars—Ribs—Cap strip—Nose strip—Stringers—Sidewalk—Struts—Wire Bracing—Wing covering—Dope—Inspection windows—Stay wires and terminal splices—Aircraft wire—Strand—Aircraft cord or cable—Terminals and splices—Soldering—Turnbuckles—Locking devices.

RIGGING deals with the erection, alignment, adjustment, repair and care of airplanes.

Airplanes are of light skeleton construction with parts largely held together with adjustable tie wires, hence they easily can be distorted or their adjustment ruined by careless or improper rigging. The efficiency, controllability, general airworthiness and safety of machine and pilot therefore depend very largely upon the skill and conscientiousness of the rigger.

For purposes of description the airplane may be divided roughly into three parts (exclusive of the power plant). These are the body or fuselage, the wings or aerofoils and the landing gear.

The fuselage is the main structural unit of the airplane. It provides a support and housing for the power plant, contains the cockpit for the pilot, and the instruments and control mechanism. The rear end of the fuselage carries the rudder, elevators, stabilizing fins and the tail skid. The wings or aerofoils are attached to the fuselage through suitable hinged connections or brackets and the fuselage is supported by the wings when the machine is in the air. Conversely the wings are supported from the fuselage when the airplane is on the ground, as in that case the whole weight of the machine is supported by the landing gear and the tail skid, both of which are attached under the fuselage.

The body or fuselage is of trussed construction, a form which gives great strength and rigidity for a given weight of material. Parts assembled together in the form of a truss are spoken of as members. Those which take a thrust only are called compression members, while those resisting a pull are known as tension members.

Other members may be either tension or compression members, depending on how the load or force is applied to them at any given time. There are also members subject to a shearing stress and others to cross-bending or compound stresses.

The fuselage is usually constructed with four main longitudinal members running the full length. These are called longerons. They are separated at intervals by compression members termed struts. The whole structure is in turn tied together and braced by means of diagonal wires, fitted with turnbuckles for adjustment, which go under the general name of wire bracing or stay wires.

Stay wires in certain parts of an airplane are designated as flying, ground, drift, anti-drift, etc. These will be considered later.

That part of the surface of the fuselage which is bounded by two struts and two of the longerons is known as a panel. The points at which the struts join the longerons are called panel points or stations. The cubical space enclosed by eight struts and the four longerons is called a bay. Some makers, Curtiss for instance, number the stations in the fuselage from front to rear calling the extreme front station No. 1. Others, such as the Standard, number these stations from the rear toward the front, calling the tail post zero.

The longerons are made of well-seasoned, straight-grained ash. They are curved inward toward the front end and usually terminate in a stamped steel nose plate. This is true particularly of airplanes equipped with engines of the revolving cylinder type. The nose plate is stamped from plate steel about .10 in. in thickness. This plate not only ties the longerons together at the front end of the fuselage, but supports one end of the sills on which the engine rests. In some types of planes it also forms a bracket for supporting the radiator. In other types of airplanes the longerons may terminate at the front end of the fuselage in an open frame which forms the support for the radiator and also supports the front ends of the engine bearers or sills. The two upper and the two lower longerons are brought together in pairs one above the other at the rear end of the fuselage, and are joined to the tail post or vertical hinge post on which the rudder is mounted.

Lightened Construction

In order to lighten the construction of the fuselage as much as possible, the rear portions of the longerons are often cut out to an I section and spruce is often substituted for ash for the rear half, suitable splices strengthened with fish plates being used wherever joints are made in the longerons. It is possible to lighten the rear portion of the fuselage in this way for the reason that this part of the body does not support as much weight or undergo as severe stresses as the forward portion.

In a machine of neutral tail lift (one in which the rear horizontal stabilizers are set at such an angle that they barely sustain the weight of the rear portion of the machine when flying horizontally in the air) the stresses in the longerons are exactly the opposite when the machine is in the air to those obtaining on the ground. When the machine is at rest on the ground it is supported near the front and rear ends of the fuselage by the landing gear and the tail skid. This method of support produces tension in the lower longerons and compression in the upper.

When in the air the machine is supported by the wings which are attached to the fuselage at the center wing section. The system of supports, trusses and stay wires between the upper and lower wings transfers most of the support from the wings to the center panel section of the upper wing. This results in tension in the upper longerons and compression in the lower.

The fuselage struts are usually made of spruce, although ash is sometimes used. The struts are joined to the longerons by means of metal clips. The construction of the clips, which are usually bent in U shape, is such that each forms a partial socket for receiving the end of a strut or struts. In general, struts are subjected to compression only. For this reason spruce is the favorite wood for struts as it is very strong along the grain in tension or compression. The strength of steel, weight for weight, would have to be 180,000 lbs. per square inch to equal spruce for this purpose. Spruce is not, however, very strong across the grain and splits readily, hence it is not a great favorite for parts subject to shearing or cross-bending stresses. On account of the liability of spruce to splitting, the ends of the struts are sometimes encased in copper ferrules or bands. This prevents crushing, splitting and chafing.

Compression Struts

When a member is subjected to a compression force it tends to bend or buckle in the center. To resist this tendency, struts subject to compression stress are made larger in the center than at the ends.

Ash is selected for the longerons because it is strong for its weight (about 38 lbs. per cu. ft.), very elastic and can be obtained in long, straight-grained pieces free from defects. It is strong across the grain so that it is able to resist the compression due to clips and struts attached at various points on the longerons.

The metal clips in which the ends of the struts are mounted are punched from sheet steel then pressed to form. They are frequently made of two or three separate pieces which are then electrically spot-welded together. They are made of .28 to .30 percent carbon steel.

The lower cross-members of the fuselage at stations 3 and 4, numbered from the front, terminate in a half hinge to which the lower wing sections are attached on either side of the fuselage. These cross-members serve as compression members when a machine is on the ground, but when it is in the air they become tension members.

Engine Bearers

The engine bearers are made of spruce with a strip of ash glued on top and bottom. They are further protected against crushing, at points where the engine supporting arms rest on the sills or stringers, by means of a copper band.

There is usually a fire screen between the engine space and the cockpit. This is to prevent injury to the pilot so far as possible in case of a back fire or fire in the engine space.

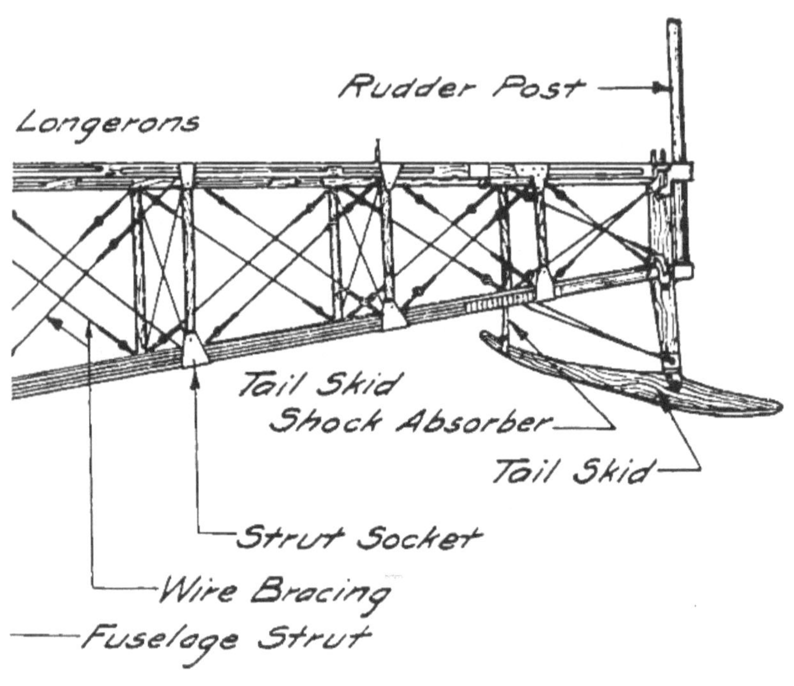
Rudder Post
Longerons
Tail Skid
Shock Absorber
Tail Skid
Strut Socket
Wire Bracing
Fuselage Strut

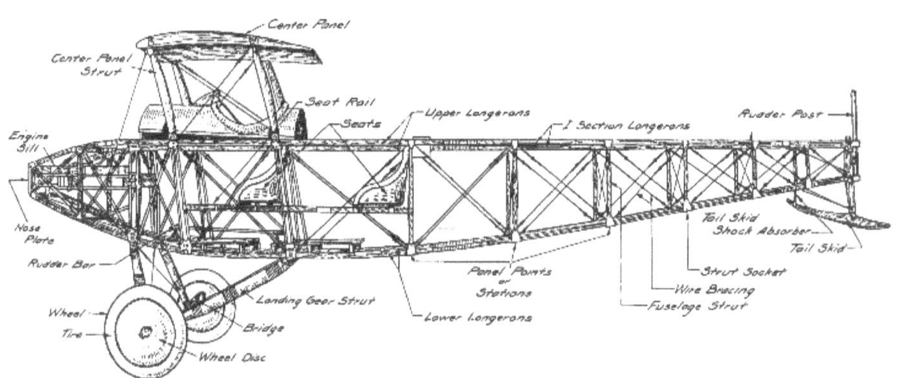

Fig. 31—Showing principal parts of fuselage

The seat rails are short longitudinal members forming supports for the pilot's and observer's seats. These rails, which are mounted on either side of the fuselage, are attached to adjacent vertical struts at the proper distance above the lower longerons.

The rudder bar is a cross bar pivoted at its center and mounted a short distance above the floor of the fuselage. It is used to control the vertical rudder and is operated by the pilot's feet. Ordinarily the ends of the rudder bar project through the sides of the fuselage, working in suitable slots cut for them, and the rudder wires are attached to the ends of the rudder bar outside of the fuselage. In machines fitted with dual controls there are, of course, two rudder bars and these are fastened together by means of wires connecting their outer ends. The rear of the two rudder bars is then connected to the vertical rudder in the usual way.

Wing section struts are vertical or diagonal struts mounted above the fuselage and attached by means of strut sockets to the upper longerons. The wing section struts are used to support the center wing panel when the machine is on the ground and when in the air they help to support the fuselage from the center panel, the latter being supported partly by the upper wing sections which are attached on either side of it and partly by the lower wing sections which are braced to the upper sections and also attached on either side of the fuselage as previously described.

The strut sockets in which the lower ends of the wing section struts are mounted consist of U-shaped steel plates firmly attached to the upper longeron. The wing section struts are mounted between the side walls of the socket, usually by means of a heavy through-bolt.

Standard Fuselage Construction

The type of fuselage just described, which is of wood and metal construction, may be said to represent standard practice in this country at the present time. There are, however, other types of construction, such as the all-steel fuselage. In this the shape of the members and the methods of joining them follow closely standard methods in structural steel work. It is claimed for the all steel construction that it is lighter for a given size machine than the wood and metal or composite construction.

The fuselage is usually covered either with canvas or linen material similar to that used for wing coverings or else with very thin panels of veneered wood. In the former case the longerons, struts and braces must carry all the weight and take up all the stresses to which the fuselage is subjected, but when a veneered wood covering is used, it contributes materially to the strength of the fuselage, consequently the framework of the latter may be made lighter.

There are also fuselages of the monocoque type in which the strength is obtained not by a truss construction, but by the form and nature of the outer shell itself, this being made up of alternate layers of thin wood veneering and cloth until the desired thickness and strength are obtained.

The various layers of wood veneering are laid with the grain running in different directions in the different layers. This type of shell or body, which is usually somewhat fish-shaped, possesses the necessary strength and elasticity without the system of struts and tie wires common to the ordinary or trussed type of fuselage. The monocoque construction possesses one marked disadvantage, however, and that is that it is very hard to repair in case of slight damage.

It may be added that the monocoque or laminated wood construction is far more common in foreign countries, particularly France and Germany, than in the United States.

Landing Gear

The landing gear is an assembly of struts, fittings, axle, wheels, shock absorbers and bracing wires whose function is to enable the machine to rise from and land on the ground and to furnish the main support of the machine when resting on the ground.

The struts of the landing gear are of streamline shape to reduce the resistance when flying. They are usually made of well-seasoned, straight-grained ash or spruce. Very often they are further strengthened by several wrappings of linen twine. The struts with their fittings constitute important members and should be carefully examined at frequent intervals. Failure or collapse of these struts would be almost certain to cause a serious accident when landing.

These struts are attached to the lower side of the fuselage, usually to the lower longerons themselves by means of metal socket fittings. The lower ends of the struts on each side of the landing gear are joined together by a metal bridge. This bridge not only serves to tie the lower ends of the struts together, but it also forms a yoke or housing in which the axle box plays up and down. The bridge is made of a steel stamping or drop forging.

The axle box may be in the form of a whole box or a half box. When it is in the form of a half box it is generally called a saddle. Its purpose is to support the axle and to guide its vertical motion in the bridge. The saddle may be either of bronze or aluminum. It is held in its place in the bridge by a wrapping of elastic cord, which consists of a number of strands or bands of rubber bunched together and enclosed in a loosely-braided covering.

The assembly of the saddle, bridge and elastic cords is called the shock absorber.

The axle is made of steel tubing and is enclosed, between the bridges connecting the pairs of struts, in an axle casing. This is made of wood, or sheet metal, built around the axle itself and is of streamline shape or section to reduce air resistance.

The wheels are the ordinary type of wire wheels of rather small diameter and usually fitted with pneumatic tires. They do not, however, ordinarily run on ball bearings, as a slight amount of friction in the wheel bearings is of little or no consequence when leaving the ground at the commencement of a flight, and it assists somewhat in bringing the machine to a stop without going too far after alighting. The sides of the wheels are covered with linen cloth discs to decrease air resistance.

Not all landing gears are like the one described, but this may be taken as standard practice. Some are provided with a skid or a single wheel projecting ahead of and above the main wheels for the purpose of preventing the machine from taking a header or nosing into the ground on landing, in case it strikes the ground at too sharp an angle. Other minor details of construction will be noted, too, on different types of machines, particularly in the construction of the shock absorbers.

The tail skid is a skid or arm projecting below the fuselage near its rear end. The purpose of the tail skid is twofold; first, to support the rear end of the airplane when on the ground or in landing and prevent damage to the rudder and elevators and their controls, and secondly, to act as a drag or brake to assist in bringing the machine to a stop when landing. The tail skid is frequently hinged or pivoted where it is attached to the lower longerons and its upper end, extending above the pivotal point, fitted with rubber cords similar to those used in the shock absorbers on the axle of the landing gear. This construction acts the same way as the shock absorber and prevents damage to the empannage and rear portion of the fuselage when landing.

Airplanes are often fitted with wing skids which consist of small auxiliary skids under the outer ends of each lower wing. These skids ordinarily do not come into action and are only provided to prevent damage to the outer wings in alighting on rough ground or in case a sudden side gust of wind should tend to upset the machine when alighting or rising.

Landing Gear of Seaplanes

Seaplanes and flying boats are of course fitted with entirely different types of landing gear from that described. Seaplanes are fitted with pontoons or floats suitable for arising from and alighting on the water. Usually there are one or two main pontoons under the forward section of the fuselage, these corresponding roughly to the main landing gear of the airplane. There is also a smaller pontoon mounted under the rear end of the fuselage and one under the outer end of each wing to prevent the wings dipping or the whole machine upsetting in rough water. The flying boat is so constructed that the whole fuselage is in the shape of a boat and the whole machine is therefore supported on the fuselage when resting on the water and when alighting and rising from the water. The flying boat is also usually fitted with small auxiliary pontoons under the outer end of the wings to keep the machine steady in rough water.

The main members running the full length of the wing are called the spars. They are usually spoken of as front and rear spars. Sometimes the front spar is called the main spar.

The cross members joining the spars together are called ribs. There are two kinds of these, compression ribs and the web ribs. The function

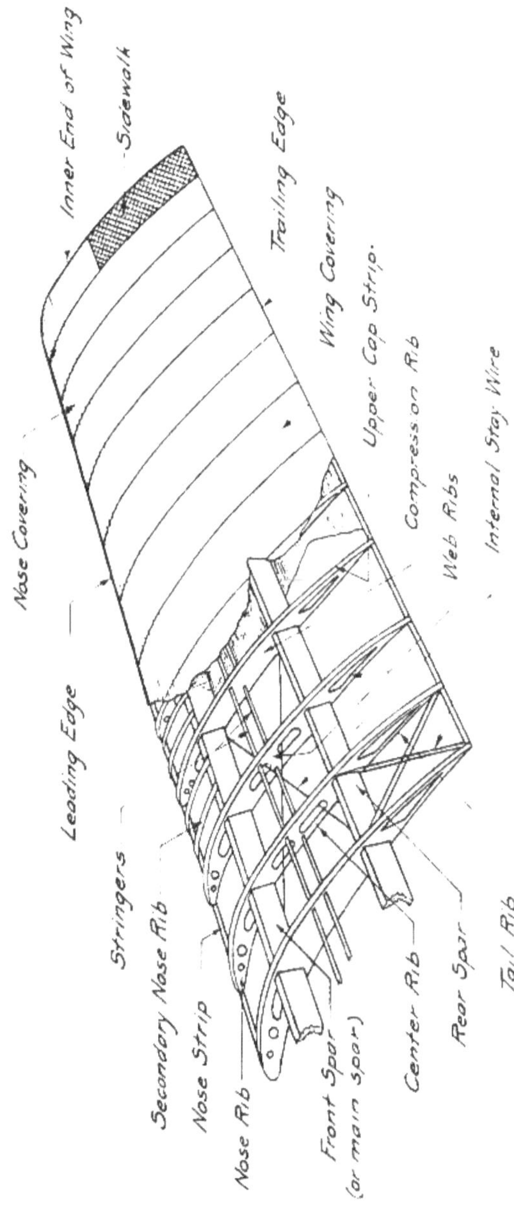

Fig. 32—Details of wing construction

of the web ribs is merely to support the linen covering of the wings and to resist the lifting force of the air, due to the forward motion of the airplane. There is not much end pressure against these ribs, therefore, the central portion is cut out for the sake of lightening them. The function of the compression ribs is not only to resist the lifting force of the air, but also to take the thrust due to the stay wires.

The ribs are not continuous, that is, they do not pass through the spars. The ribs are made in three sections, the nose section, center section and tail section. The nose section of a rib is the section which projects forward of the front or main spar. The center section is the section between the front and rear spars. The tail section of the rib is that which projects to the rear of the rear spar. The nose sections and tail sections are sometimes called nose ribs and tail ribs and are also frequently spoken of as nose webs and tail webs, because they are cut out to a web form. These rib sections are not, of course, called upon to stand compression stresses, as these stresses are all centered in or taken through the front and rear spars.

A thin strip of wood running from the nose web across the spars to the rear end of the tail webs (lengthwise of the airplane itself) and serving to bind all the wing parts or ribs together, is called the cap strip. There is a top cap strip and a bottom cap strip on each set of ribs.

Entering and Trailing Edges

The front edge of the wing section which is the part carrying the nose webs or nose ribs is called the entering edge of the wing. The rear edge of the wing is known as the trailing edge.

The nose webs are tied together by a strip of spruce running full length of the wing or crosswise of the airplane itself. This strip forms the leading edge of the wing and is called the nose strip. From the nose strip to the front or main spar, on the upper side of the wing, there is a covering of thin laminated wood called the nose covering. Its purpose is to reinforce the covering fabric as it is at this point that the effect of wind pressure due to velocity is most severe.

Secondary nose ribs are placed between each pair of full ribs to give additional support to the nose covering.

There are usually two rod-like members running from end to end of the wing through the central part of the ribs. These are called stringers and are used for the purpose of giving lateral stiffness to the ribs.

The trailing edge of the wing is made of thin flattened steel tubing attached to the tail webs by metal clips.

The spars are continuous throughout their length. Furthermore, they have reinforcements of wood at the points where the interplane struts connecting the upper and lower wings are attached. Steel bearing plates are bolted to the wing spars at these points. The bolts attaching these bearing plates to the wing spars do not pass through the spars themselves, but through the reinforcements. This is to avoid weakening the spars.

Nearly all wood used in wing construction is spruce, with the exception of the nose covering which is made of birch or gum wood, the web ribs, which are made of laminated wood, and small quantities of pine or other woods in the sidewalk and other unimportant places.

The sidewalk is a boxed-in or wood-covered portion of the inner end of the lower wing. It furnishes a solid footing for the pilot or observer when entering or leaving the cockpit and for mechanics working around the engine, guns, instruments, control mechanism, etc.

Steel hinge pieces are bolted to the inner ends of the wing spars and serve as a means of connecting the lower wings to the fuselage and the upper wings to the center wing panel.

Interplane struts are vertical or inclined wooden struts of streamline section used to transfer compression stresses from the lower wings to the upper wings when the machine is in flight. These struts are used in conjunction with diagonal stay wires which serve to transfer the load towards the center of the machine when in flight.

The stay wires are divided into two general groups, those which take the drift load or fore-and-aft stresses due to the forward motion of the airplane, and those which take the lift load or vertical load due to the weight of the machine itself and the vertical resistance when in the air. The lift wires are again divided into those which take the load when the machine is flying and those which take it when on the ground. The wires which take the lift load when the machine is in the air are called the flying wires, and those which take the load when on the ground are called ground or landing wires.

Drift and Anti-Drift Wires

The set of wires in the wings which carry the drift load when flying are called the flying drift wires, or drift wires for short. There is no reversal of load in these wires when the machine is on the ground, but opposition wires are necessary to maintain structural symmetry. These latter are called the anti-drift wires.

When the wing frames are covered it is of course impossible to inspect the internal stay wires of the wings, hence every precaution must be taken to guard against corrosion. The wire used at this point is tin coated before assembling, the steel parts of the turnbuckles and other fittings are copper plated and when completely assembled, all the metal parts are given a coat of enamel paint. All screws, tacks and brads are of brass or copper.

Wings are covered with a closely woven fabric. At present unbleached linen seems to give the best satisfaction. Owing to its scarcity, however, a satisfactory substitute is being sought for. A cloth made of long fibre sea island cotton is used to some extent and makes a fairly satisfactory substitute.

Linen fabric weighs $3\frac{1}{2}$ to $4\frac{3}{4}$ oz. per sq. yd. and has a strength of 60 to 100 lbs. per in. of width. Its strength is increased 25 to 30 percent by doping, however. The weight of cotton fabric is 2 to 4 oz. per sq. yd., its strength 30 to 60 lbs. per in. of width, and its strength is increased 20 to 25 percent by the application of dope.

The cloth surfaces or wing coverings must be taut, otherwise on passing through the air they would vibrate or whip. This would not only increase the resistance to a great extent, but soon would lead to the destruction of the fabric. A preparation called dope is used to tighten up the fabric and give a smooth, taut surface. It also tends to make the cloth weather-proof.

Dope should be easy of application, durable, fire resisting and have a preserving effect on the cloth. Dopes at present are divided into two classes or chemical groups, those which are made from a base of cellulose nitrate or pyroxylin and those made from a cellulose acetate base. The base is dissolved in a suitable solvent, such as acetone for instance, and sometimes other substances are added to preserve flexibility or prevent drying out and cracking and checking or to modify shrinkage.

The greatest difference between these two dopes is in their relative inflammability. The acetate dope makes the fabric not fireproof, but slow burning. A cloth treated with this dope will shrivel and char before burning, but one treated with nitrate dope will burst into flame immediately on the application of a lighted match or when exposed to a strong spark or punctured by a flaming bullet, etc. See "Airplane Dopes," by Gustavus J. Esselen, Jr., in *Aviation*, July 5, 1917.

Inspection windows are often inserted in wing sections over and under certain control joints where the latter are carried inside the wing section itself. For instance, the aileron control cables are frequently run inside the lower wing sections to a pulley attached to the front or main spar opposite the middle of the aileron, the cable then passing down at a slight angle and through a thimble or sleeve in the lower covering of the wing section to the point where the cable is attached to the aileron control mast. With this construction inspection windows would be set in the upper and lower coverings of the lower wing immediately above and below the pulley over which the control cable passes. The inspection windows are usually of celluloid or other transparent material firmly sewn into the wing covering material.

Stay Wires and Splices

Stay wires and cables are used extensively in airplane construction. Much of the safety of the machine and pilot depends upon the quality of the material in the stay wires, the care used in adjusting them and on the character of the terminal splices.

Three kinds of materials are used for stay wires: solid or aircraft wire, stranded wire or aircraft strand, and a number of strands twisted together to form a cable and known as aircraft cord. Aircraft wire is a hard drawn carbon steel wire coated with tin to protect it against corrosion. Its strength runs from 200,000 to 300,000 lbs. per sq. in., depending upon how small it is drawn. Drawing increases both the strength and hardness of this type of wire, but if drawn until too hard it cannot be bent with safety. The aim is to produce a wire of maximum strength, yet with sufficient toughness to allow it to bend without fracture. A standard test for bending is to grip the wire in a vice whose jaws have been rounded off to 3/16 in. radius, and bend the wire back and forth through an angle of 180 deg. Each bend of 90 deg. counts as one bend. The minimum number of bends for various sizes of aircraft wires should be as follows:

For wire of B. & S. gauge No. 6— 5 bends without fracture.
For wire of B. & S. gauge No. 8— 8 bends without fracture.
For wire of B. & S. gauge No. 10—11 bends without fracture.
For wire of B. & S. gauge No. 12—17 bends without fracture.
For wire of B. & S. gauge No. 14—25 bends without fracture.
For wire of B. & S. gauge No. 16—31 bends without fracture.

Aircraft strand is composed of a number of small wires, usually 19, twisted together. The individual wires of the strand are galvanized or zinc coated before being twisted into the strand. The complete strand

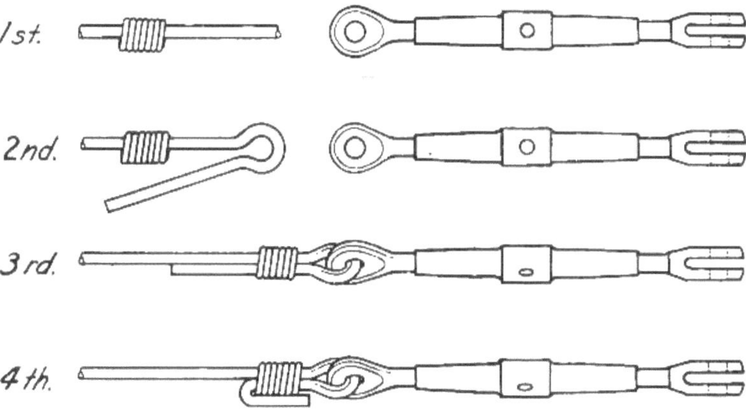

Fig. 33—Steps in making an end splice in solid wire

is more flexible than a solid wire of the same diameter and is therefore more suitable for stay wires that are subject to vibration.

The stay wires of the fuselage at the engine and wing panels are of aircraft strand or cord, but for the remaining stay wires of the fuselage aircraft wire is ordinarily used.

Aircraft cord is much more flexible than the strand. It is used for control cables where these must pass over comparatively small pulleys. The usual construction of aircraft cord is 7 strands of 19 wires each twisted together to form a cable. This specification is known as 7 x 19 aircraft cord. The individual wires of the cord are very small and are tin-plated before being stranded.

For a given diameter, the solid wire is stronger than either the strand or cord. Weight for weight, however, the cord is a little stronger than the wire, as shown by the following table.

	Weight per 100 ft.	Strength for a given diameter	Strength for a given weight
Wire	8.84 lbs.	5500 lbs.	5500 lbs.
Cord	6.47 lbs.	4200 lbs.	5600 lbs.

A wire or cord is no stronger than its terminal splice. The splice may be formed in a variety of ways. For solid wire the formation of the eye is important. An eye in which the reverse curve has the same radius as the eye proper is called a perfect eye and is the one recommended. The inside diameter of the eye should be about three times the diameter of the wire itself.

After the eye is formed a flattened wrapped wire ferrule, somewhat like a coiled spring flattened to eliptical section, is slipped over the wire and the free end. The latter is then bent back over the ferrule. Such a terminal will have an efficiency of 60 to 65 percent of the strength of the wire itself. When this type of terminal fails it is usually by slipping. If the free end of the wire is tied down, after being bent back over the ferrule, with an additional wrapping of wire, the efficiency of the terminal as a whole will be increased to 80 percent of the strength of the wire. If the whole terminal is soldered the efficiency will be increased to 100 percent according to static tests. This is misleading, however, as such tests take no account of live load stresses or vibration.

Another form of terminal is made by substituting a thin metal ferrule or section of flattened tube for the wrapped wire ferrule. It can be made secure either by soldering or twisting after being put in place. This terminal for live or vibrational loads is superior to the wrapped wire terminal as there is not so much difference in mass between the wire and the ferrule.

Aircraft Strand Terminals

The terminal eye of the aircraft strand is formed around a thimble. The free end of the strand is brought around the thimble and either wrapped to the main strand with small wires and soldered, or the free end is spliced into the main strand. Before bending around the thimble, the strand is wrapped with fine wire in order to prevent flattening or caging of the strand.

The terminal eye of the aircraft cord is always made by splicing the free end of the cord into the main strands after wrapping the cord around a thimble. Sometimes the splice is soldered but more often it is wrapped with harness twine. Foreign engineers are opposed to soldering, claiming that the disadvantages in the way of corrosion and overheating of the wire outweigh the advantages of the stronger terminals.

The theory of the splice is simple. A strand or wire of the free end is wrapped around a strand or wire of the main cord, care being taken to have the lay of the wires the same. Three to five complete turns are given, three for the first and four to five for the last weaves of the splice in order to taper the splice gradually.

Objections to soldering.—The most serious objections to soldering are: a. overheating; b. corrosive action of fluxes. It is very easy to overheat and soften the wire and this is all the more serious because the softening takes place at a point where the wire is enlarged by the joint. The stress is naturally localized at this point.

Some of the so-called non-corrosive fluxes will upon application be found to be more or less corrosive. Even with strictly non-corrosive fluxes, there is a carbonaceous residue, due to heat, driven into the interstices between the wires of strands or cords. This serves as a holder for moisture and will in time cause corrosion.

The corrosive effects of acid fluxes can be neutralized by the application of an alkaline solution, such as soda water. Washing the soldered splice of a solid wire with such a solution is very effective, but with strands and cords, where the acid is driven into the interior through the application of heat, it is questionable whether any system of washing will eliminate or neutralize the acid. Corrosion of the interior wires of a strand or cord may be concealed by a perfectly good exterior, giving an entirely false appearance of security.

Turnbuckles

Turnbuckles are made of three parts, the ferrule or sleeve, and the two ends. To distinguish the ends, they are called the yoke and eye ends, or the male and female.

Great care should be exercised when tightening or loosening turnbuckles that the cables are not untwisted or frayed. If the cables are untwisted a caging of the strands results which greatly weakens the cable. Cable that has been caged should be replaced. No pliers should be used when tightening or loosening turnbuckles. The correct method is to use two drift pins or nails, one through the terminal eye of the cable to prevent the end of the cable twisting, the other through the hole in the barrel of the turnbuckle. Pliers will scar the wires, which is objectionable for three reasons, the first two of which may lead to serious consequences. These reasons are: First, breaking the protective coating given to guard against corrosion. Second, a nick or scar in a wire or cable which would weaken it considerably. The wire or cable may not show much reduction of strength under a static load or test, but with a live or vibrational load the strength is greatly reduced and a slight nick will determine the point of fracture. Third, disfiguration of the parts is offensive to the eye and bespeaks slouchy or careless workmanship.

Locking Devices

A fair proportion of accidents occurs to moving mechanism through nuts or other threaded fastenings working loose. It is safe to say that several hundred patents have been taken out for nut-locking devices, but of this great number, a few only are of practical value and used to any extent. The castellated nut and cotter pin used of course with a drilled bolt or stud is one of the few devices that finds large application. It is generally used in automobile and airplane work. The spring locking washer is another good device. This is used where the fastening is of a permanent or semi-permanent character. Another method is to batter or hammer down the end of a bolt a little. This should be practiced only as a last resort or as an absolutely permanent job and must be carefully done, otherwise serious damage will result to the bolt and nut. It is sufficient to close one thread on the bolt for part of the circumference only.

Turnbuckles are secured against turning or loosening by running a wire through the adjusting hole in the turnbuckle sleeve and carrying the wire back and binding it around the ends of the turnbuckle. See Fig. 41.

Chapter 5

ALIGNMENT

Fuselage alignment—Horizontal and vertical stabilizers—Landing gear or under-carriage—Center wing section—Wings—Lateral dihedral angle—Table for lateral dihedral—Stagger Overhang—Rigger's angle of incidence—Wash-out and wash-in—Overall measurements—Aileron controls—Elevator controls—Rudder control Notes on aligning boards.

BY the term airplane alignment is meant the art of truing up an airplane, and adjusting the parts in their proper relation to each other as designated in the airplane's specifications. The inherent stability, the speed, the rate of climb, the efficiency, in short the airworthiness of an aircraft depend in large measure on its correct alignment. For this reason the importance of careful and correct alignment cannot be overestimated.

The instructions as given in this chapter are not intended to be a complete and exhaustive treatise on the whole subject of airplane alignment, but are designed rather to give the beginner a good general idea of how the work is done. Thus with these instructions as a ground work he can become proficient in the work after having had good practical experience in the hangars.

The work of aligning an airplane divides naturally into several distinct and separate groups or divisions—a. fuselage, b. horizontal and vertical stabilizers, c. landing gear, d. center wing section, e. wings, f. controls.

Alignment of fuselage.—The fuselage is aligned before leaving the airplane factory and normally this alignment will last for some time. The fuselage alignment should be checked over carefully, however, after an airplane has been shipped in disassembled condition. Strains on the fuselage caused by rough handling, bad landings, etc., will make it necessary to re-align it.

Before attempting to align any part of an airplane the erection drawings should be referred to if available, and the directions furnished by the makers should be followed carefully unless the operator has had a great deal of previous experience upon the particular type of airplane to be aligned, and is familiar with better methods of procedure than those recommended by the maker.

In general the procedure in aligning a fuselage will be about as follows: A horizontal reference plane is usually specified by the makers in connection with the fuselage. Sometimes the top longerons are taken as this reference plane, in which case they are to be aligned horizontally, laterally, and longitudinally from a specified station to the tail post. Sometimes horizontal lines are drawn on the vertical fuselage struts, and the fuselage is so aligned that these lines all fall in the same horizontal plane.

Alignment of Longerons

In the first case, after the fuselage has been placed in a flying position, the top longerons are aligned for straightness, using a straight edge and a spirit level to aid in finally placing them laterally and longitudinally in a horizontal plane.

The longerons are next aligned symmetrically with respect to the imaginary vertical plane of symmetry through the fore-and-aft axis of the fuselage. There are two general methods of doing this, as follows:

First Method—The center points are marked on all horizontal fuselage struts. A small, stout cord is stretched from the center of the fuselage nose to the tail post and the horizontal bracing wires adjusted until the centers of the horizontal struts fall beneath this line. A small surveyor's plumb bob is held at different points so that the suspending cord just touches the fore-and-aft aligning cord. The centers of the bottom horizontal struts should fall directly below the bob.

Second Method—A plumb line is dropped from the center of the propeller and from the tail post and a string is stretched on the ground or floor between these two points. Plumb bobs dropped from the centers of the horizontal struts must point to this line.

The whole fuselage alignment is checked to make sure that it agrees with the specifications. If the airplane has a non-lifting tail, it would be advisable as the next step to support the fuselage in such a way that the rear part (about two-thirds of the total fuselage length) remains unsupported, and then re-check the fuselage alignment once more.

All turnbuckles should then be securely locked and the fuselage carefully inspected.

Horizontal and Vertical Stabilizers

The vertical stabilizer is examined to see that the bolts holding it in place are properly drilled and cotter-pinned, also to see that it is set parallel or dead on to the direction of motion. It is trued up vertically by the turnbuckles on the tie wires or brace wires connected to it. These turnbuckles are then properly safetied.

The horizontal stabilizer usually is braced with tie wires fitted with turnbuckles. By means of these its trailing edge should be made straight and at right angles to the horizontal center line of the fuselage. All bolts fastening the horizontal stabilizer to the fuselage should be inspected to make sure they are properly drilled and cotter-pinned. All turnbuckles should be safetied, as shown in Fig. 41.

Alignment of landing gear or under-carriage.—In assembling an airplane which has been completely dismantled, the landing gear should be assembled to the fuselage and aligned with it before the wings are attached. In assembling and aligning the landing gear, the fuselage should be so supported that the landing gear hangs free and the wheels do not touch the ground.

The fuselage is placed in the flying position, or at least in such a position that the lateral axis is horizontal. There are three general methods of aligning the landing gear, as follows:

First Method—A small plumb bob is dropped from a point on the fore-and-aft center line of the fuselage above the axle of the landing gear. A tack is placed in the exact center of the axle casing or a scratch is made on the axle at its center. The transverse tie wires are then adjusted until the tack or center line mark falls exactly below the plumb bob. The wires are made moderately tight. The exact degree of tautness required cannot very well be described; it is a matter of experience or personal instruction. All turnbuckles are safetied and the landing gear inspected carefully. The strut fittings and the elastic shock absorbers should be inspected very carefully.

Second Method—The two forward transverse tie wires are adjusted until equal in length, then the rear transverse tie wires are similarly adjusted until they also are equal in length. All transverse tie wires are tightened equally and the turnbuckles safetied. The landing gear is then given a final inspection.

Third Method—The transverse tie wires are adjusted until the axle is horizontal as shown by a spirit level. This adjustment is made with the fuselage in the flying position or with the lateral axis horizontal. The transverse tie wires are tightened equally to the correct tautness, the turnbuckles safetied, and the landing gear inspected as before.

Center Wing Section

Alignment of center wing section.—The fuselage is first placed in the flying position, and the center wing section adjusted symmetrically about the fore-and-aft center line of the fuselage in plan. A tack driven in the middle of the leading edge of the center panel will then be directly above the center line of the fuselage. This is tested with a small plumb bob and checked by measuring each pair of transverse tie wires to see if the two wires of each pair are equal in length.

The alignment for stagger is made by adjusting the stagger or drift wires in the fore-and-aft direction until the leading edge of the center panel projects the required distance ahead of the leading edge of the lower plane as given in the airplane specifications. This alignment is checked by dropping a plumb bob from the leading edge of the center panel and measuring forward in a horizontal plane from the leading edge of the lower plane to the plumb line. The adjustment for stagger fixes the rigger's angle of incidence. All turnbuckles are safetied and the alignment re-checked.

Alignment of wings.—Before any attempt is made to align the wings the fuselage should be carefully inspected to make sure that it is properly rigged and in proper alignment. Failure to do this may cause much delay and waste of time in aligning the wings.

The next step is to make a general inspection of the wings, noting if all bolts and clevis pins are properly cotter-pinned. Note particularly the clevis pins where the interplane brace wires are fastened to the upper plane fittings. One of the largest airplane makers in this country puts these clevis pins in head down. In this position if the pins are not properly cottered, there is great danger of their working loose and dropping out, disconnecting the wires. Such matters are more easily remedied before the wings are aligned than afterwards.

Loosen all wires between the planes including flying wires, ground wires, stagger wires and external drift wires. Examine the turnbuckles to see that the same number of threads show at both ends. If not, take the turnbuckle apart and remedy this. It will mean a saving of time in the end if these matters are looked after before the actual truing up of the wings is begun.

Flying Position

Place the fuselage in the flying position as defined in the airplane's erection drawings. This may mean aligning the top longerons or the engine bed or other specified parts laterally and longitudinally horizontal. This must be done carefully, using a good spirit level, because the wings are aligned from the fuselage upon the assumption that this flying position is correct. If it is necessary to get into the cockpit or in any other way disturb the fuselage during the alignment of the wings, make sure that the fuselage is still in the correct flying position before proceeding further.

Lateral dihedral angle.—There are three common methods of adjusting for lateral dihedral:

Aligning Board

First Method—Aligning Board.* If an aligning board is available its use saves considerable time due to the fact that the rigger secures the lateral dihedral angle, straightness of wing spars, and correct angle of incidence near the wing tips all at the same time. The protractor level should read directly in degrees. Set this instrument at the number of degrees dihedral stated in the airplane's specifications. Place the aligning board parallel to the front spar (by measuring back from the strut fittings) and, keeping the flying and stagger wires loose, pull up on the ground wires until the bubble on the protractor level reads almost level. Since the aligning board is a straight edge it is easy to keep the front spar perfectly straight by glancing beneath the aligning board occasionally. It should rest on at least three ribs, one near each end and one near the middle. The space between the other ribs and the aligning board should be slight.

*See note on aligning boards at end of this chapter.

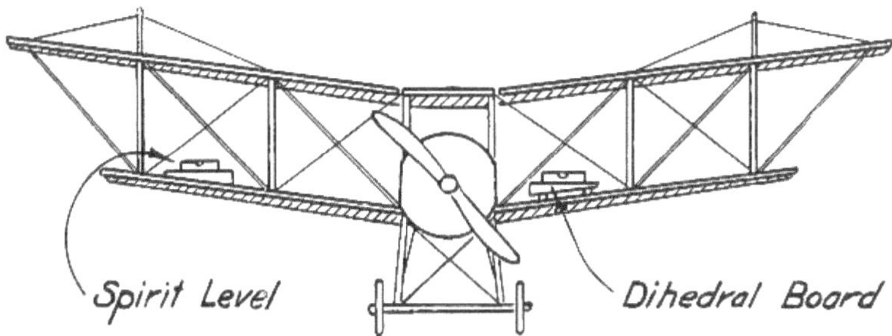

Fig. 34—Method of using short dihedral board

Place the aligning board in front of and parallel to the rear spar. Adjust the ground wires until the rear spar is straight and the dihedral is slightly greater than called for in the maker's specifications. Check at the front spar. It will now be the same as the rear. If not make it so.

Now tighten down on all flying wires except those to the overhang, if there is overhang. Test each pair of flying wires for equal tautness by striking with the edge of the hand and watching their vibration. The loose wire has the greatest amplitude of vibration. The lateral dihedral should now be exactly as called for in the specifications.

After aligning both wings for dihedral as stated above, both wings will be the same height if the fuselage is level laterally. Check the height of the wings by making the distance BA (see Fig. 35) equal to DC measured from the longerons opposite the butt ends of the front spars on the lower wing panels. V is a tack in the middle of the leading edge of the

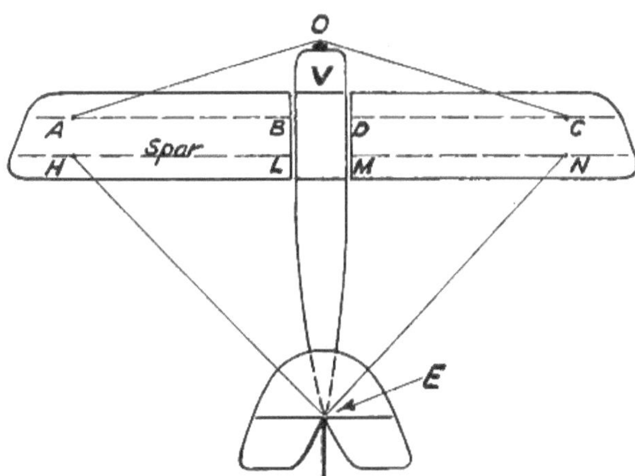

Fig. 35—Points of measurement for wing alignment

center section panel. With a steel tape measure the distance VA and VC. These distances should be equal.

Equally good results may be obtained by using a protractor spirit level in conjunction with an accurate straight edge.

Second Method—If a good aligning board is not available the string method may be used. Fig. 36 shows the arrangement of the string which should be small, smooth and tightly drawn.

Keep the stagger wires, flying wires and nose drift wires loose as in the first method. Increase the dihedral angle by tightening the ground wires, keeping the panels straight by sighting. The greater the dihedral angle the greater the distance Y (see Fig. 36). The table below shows the variation for customary range of lateral dihedral:

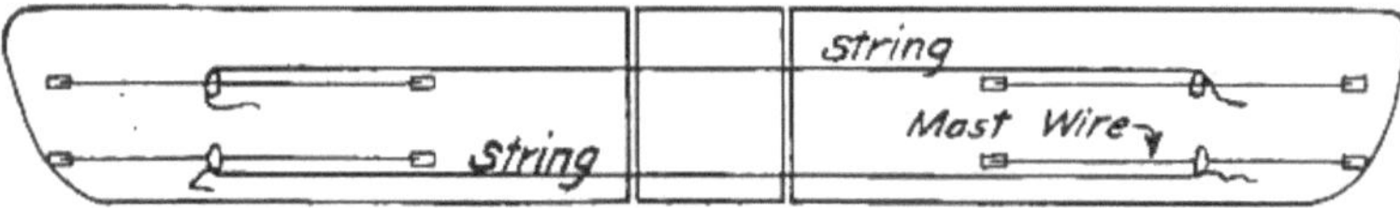

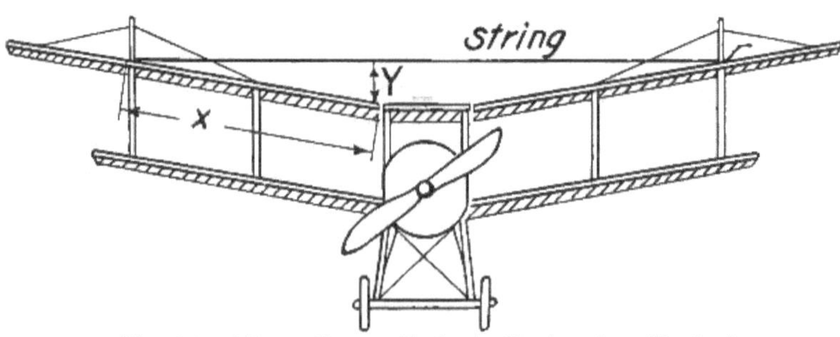

Fig. 36—Alternative method of aligning for dihedral

Table for Lateral Dihedral Angles

Degrees	X Inches Distance from point of support of string to end of spar.	Y Inches Distance from end of spar vertically up to the horizontal string.
0	100	0
1	100	$1\frac{3}{4}$
2	100	$3\frac{1}{2}$
3	100	$5\frac{1}{4}$
4	100	7
5	100	$8\frac{11}{16}$
6	100	$10\frac{7}{16}$
7	100	$12\frac{3}{16}$
8	100	$13\frac{15}{16}$
9	100	$15\frac{5}{8}$
10	100	$17\frac{3}{8}$

The distance X will probably not be exactly 100 in. as given in the table, but since X and Y increase in the same proportion this is very simple. For example, the distance X (convenient to measure) on a biplane having a 3 deg. lateral dihedral angle may be, say 12 ft. 6 in., or 150 in., which is one and one-half times 100 in.

The table gives Y equal to 5¼ in. for 3 deg. Our X is one and one-half times the X in the table. Then our Y must be one and one-half times 5¼ in. (the Y given in the table), which equals 7⅞ in., which is the proper distance up to the string when the wing has the correct lateral dihedral.

In determining the distance Y, always measure the vertical distance up to the string from near the inner edge of the wing panel, not from the center section panel. The correct lateral dihedral angle having been obtained, proceed further as in the first method.

Third Method—On airplanes having sweep-back the string method is rather difficult to apply. If an aligning board such as used in the first method is not available, then a short dihedral board may be made which will serve. Fig. 37 shows the construction and Fig. 34 the method of

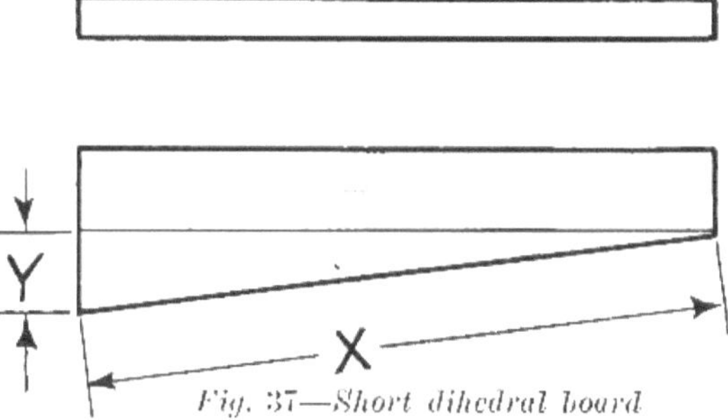

Fig. 37—Short dihedral board

using such a board. It is plain that a separate board must be made for each airplane having a different dihedral from the others at a flying field. Another disadvantage of this board is the fact that it must be used between struts on the spars and is so short that it is apt to be affected greatly by unequal rib heights and any lack of straightness in the spars.

After obtaining the correct dihedral proceed as in the first method.

Stagger is usually given in airplane specifications as a linear measurement in inches. The specifications will tell whether it is to be measured on a projection of the chord or as a horizontal distance. (See Fig. 38.) It is important to measure the stagger in the manner directed.

The stagger of the wings is fixed at the fuselage by the stagger of the center wing section. Align for stagger by adjusting the stagger wires between interplane struts. Slight adjustments only should be necessary. Fig. 38 shows the method.

In exceptional cases the flying and ground wires, front and rear, nearest the fuselage, are used in adjusting the stagger, which is usually found to be correct however after slight adjustments of the stagger wires.

Stagger is sometimes given as an angle of stagger in degrees. This can be converted into inches by the use of the lateral dihedral table on page 65. In this case AB in Fig. 38 corresponds to X in the table, and Y in Fig. 38 will be proportional to Y in the table. For instance if AB in Fig. 38 is 50 in. in a given airplane, or one-half of X in the table, and the stagger is given in the airplane's specifications as 7 deg., then the amount of stagger Y (Fig. 38) would be one-half of the 12 $^{3}/_{16}$ in. given in column Y in the table opposite 7 deg.

Overhang.—If an airplane has much overhang it is usually supported by mast wires above and flying wires below. See that the flying wires are loose. Tightening one set of wires against an opposing set throws

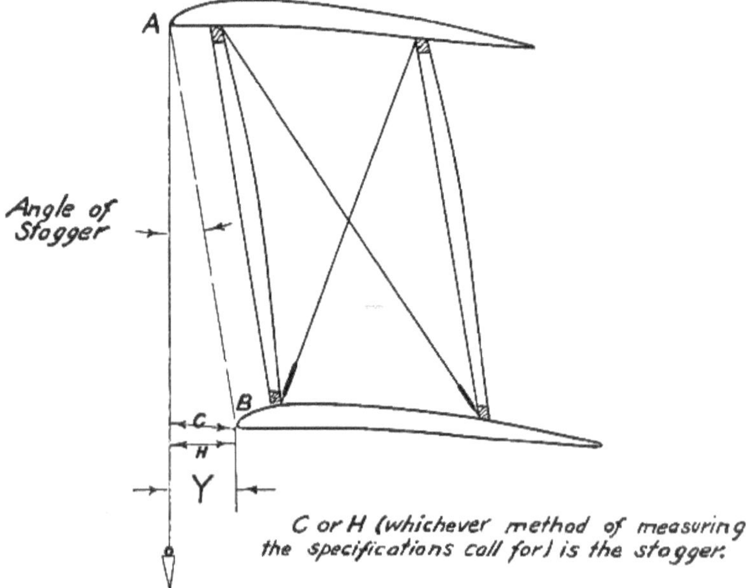

Fig. 38—Methods of measuring stagger

undue stress in members. Tighten up on the mast wires until the overhang inclines very slightly upward. Now tighten up on the flying wires below until the spars are straight.

The leading and trailing edges of all wing panels should now be straight. In case there should be small local bows in the spars, with a little careful adjusting of wires these can usually be distributed equally between the upper and lower wing panels so that their effect will be lessened. Fixing the lateral dihedral or the angle of incidence for either upper or lower plane automatically adjusts it for the other plane.

Rigger's angle of incidence.—Check the lateral dihedral to make sure that it has not been altered in making other adjustments. If it is cor-

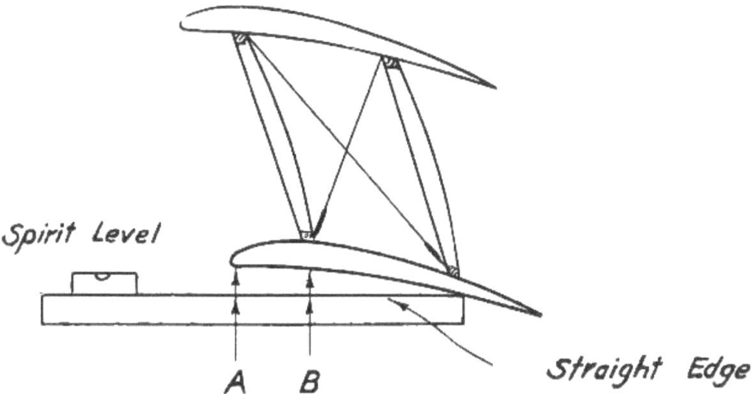

Fig. 39—Measuring angle of incidence with straight-edge and spirit level

rect, front and rear, and the spars are straight, then the angle of incidence should be correct all along the wing. Figs. 39 and 40 show two methods of testing this. If the set measurements A or B are known, the first method (Fig. 39) can be used. If the angle AOB is given in the specifications then the second method (Fig. 40) can be employed. Test the angle of incidence near the fuselage and beneath the interplane struts.

Wash-out and wash-in.—Due to the reaction from the torque of the propeller the airplane tends to rotate about its longitudinal axis. To counteract this the wing which tends to go down (sometimes referred to as the "heavy" wing) is drawn down slightly at its trailing edge towards its outer end, or in other words it is given a slight additional droop at this point. This is usually referred to as a "wash-in." The wing on the other side of the machine is given a slight upward twist, or "wash-out" at a corresponding point. In single-engined, right-hand tractors wash-

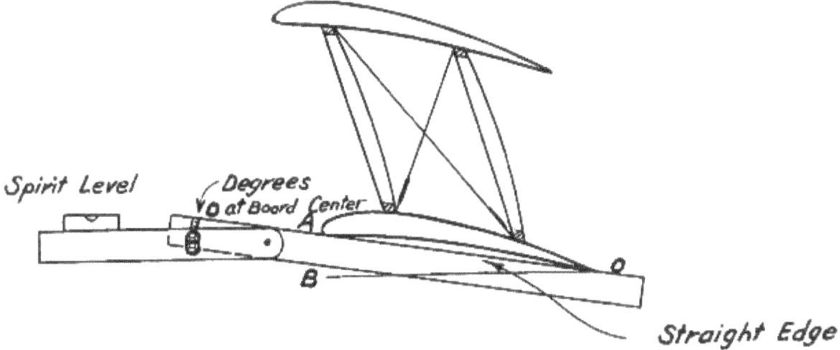

Fig. 40—Another method of measuring angle of incidence. It can also be done advantageously by using a straight edge in conjunction with a protractor spirit level

in is given to the left wing and wash-out to the right. To increase the angle of incidence the rear spar must be warped down by slackening all the wires connected to the bottom of the strut and tightening all which are connected to the top of the struts, until the desired amount of wash-in is secured. This process is reversed to secure wash-out.

For purposes of increased stability wash-out is sometimes given both wings although of course some lift is lost by doing this. If it is still desired to compensate for the reaction due to the propeller torque, more wash-out is given on one side than on the other. The side having the least wash-out then has wash-in relative to the other side.

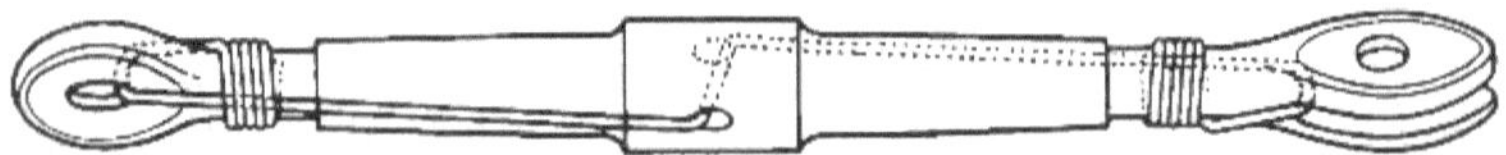

Fig. 41—The proper way to lock a turnbuckle

Over-all measurements.—Tighten the external drift wires only moderately tight. The following over-all measurements should now be taken, using a steel tape (see Fig. 35): Make BA = DC and LH = MN. Then OA should equal OC and HE should equal EN. These measurements should be made at points on the upper wing panels as well as the lower, making eight check measurements in all.

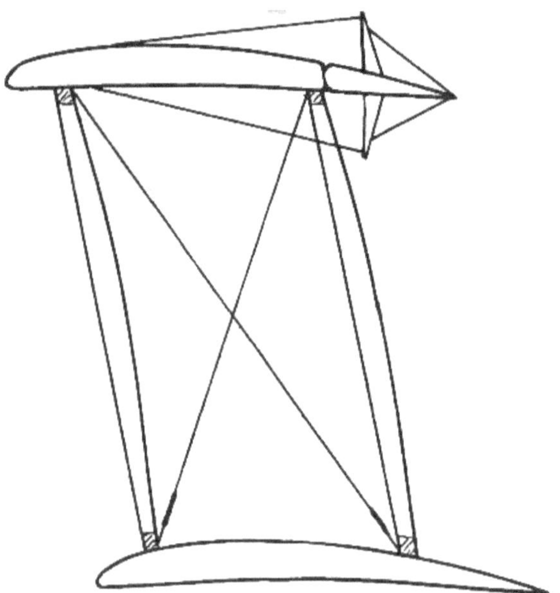

Fig. 42—Individual method of connecting aileron controls

All turnbuckles are now safetied (Fig. 41). Make a general final inspection of the wings to make sure that nothing has been overlooked. It must be remembered that in making one adjustment other adjustments made previously may be thrown slightly off, so that when the wings are finally aligned it is a good plan to check the lateral dihedral, stagger, angle of incidence, etc., to make sure that all are correct.

Controls—Ailerons.—Fasten the hand wheel, stick, or shoulder yoke controlling the ailerons in its central position. If the ailerons have brace wires on each side (see Fig. 42) and these wires are supplied with turnbuckles, straighten up the trailing edge by adjusting these wires. If the ailerons are connected as in Fig. 43 the trailing edges must be straightened as the ailerons are aligned on the airplane.

There is difference of opinion about drooping the trailing edge of ailerons below the trailing edge of the plane to which they are fastened. At some fields the turnbuckles on the aileron control cables are so adjusted that the trailing edge of the aileron lines up with the trailing edge

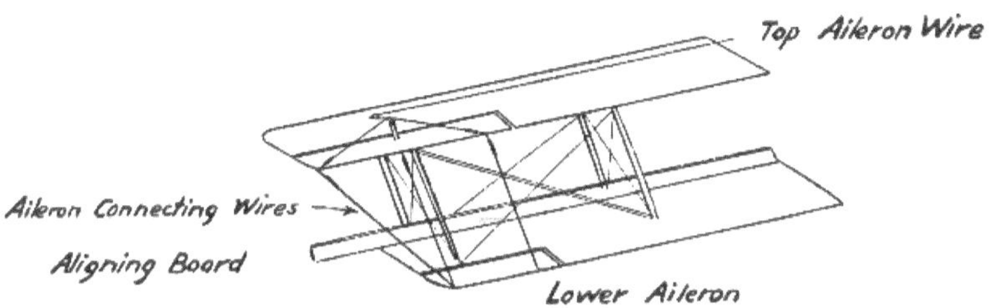

Fig. 43—Showing series method of connecting ailerons in pairs

of the wing panel to which it is hinged. At other fields, from 1/8 in. to 3/4 in. of droop is given the trailing edge of the aileron, because it forms a part of a lifting surface and it is reasoned that slack will be taken out of the lower control cables when the machine gets into the air. Unless directed otherwise it perhaps is advisable to give little or no droop.

The ailerons should work freely and respond quickly with no feeling of drag when the hand wheel is turned or the stick moved even very

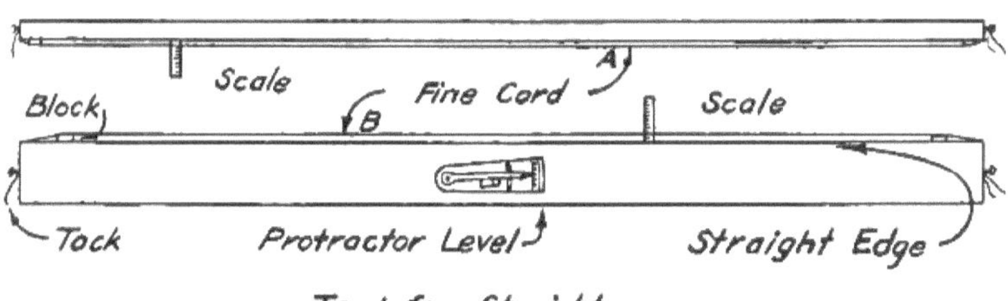

Fig. 44—Trying an aligning board for straightness

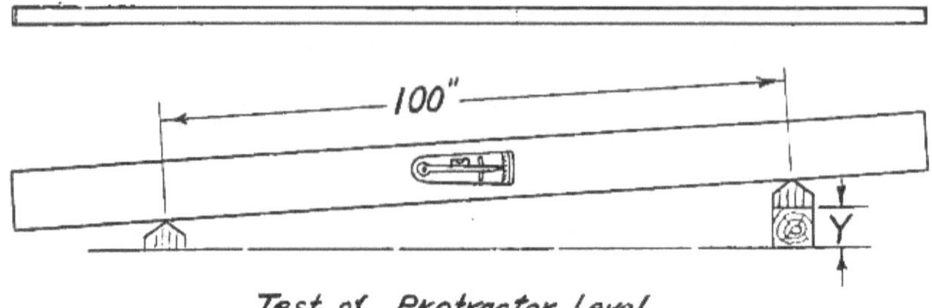

Fig. 45—Aligning board used with table for lateral dihedral angles

slightly. Improper coiling of cables when a machine is dismantled will ruin this condition about as quickly as anything could. Care must be taken not to put too much tension on the cables. The pulleys around which they run are light, and not always so strong as they might be. Cracked pulleys may sometimes be found on old machines.

Interplane ailerons are adjusted so that both are in the same plane when control is neutral. The angle at which they are set must be given by the makers or determined by experiment and experience.

Elevators.—Fasten the bridge or stick control in its central position. Adjust the turnbuckles on the control cables until the elevators are in their neutral position and both are in the same plane. Tighten the control cables enough to remove lost motion.

Rudders.—Fasten the rudder footbar in its mid-position and adjust the turnbuckles until the rudder is in the neutral position, and the cables are tight enough to remove lost motion.

Both elevators and rudders usually carry brace wires with turnbuckles which can be used in straightening their trailing edges.

Notes on Aligning Boards

To be useful an aligning board first of all must be true. Fig. 44 shows a method of testing such a board for straightness. (See A and B, Fig. 44.) Also by supporting the board as shown and setting the protractor level at different degrees the protractor can be tried out. Reference to the table for lateral dihedral on page 65 shows the difference in thickness of the blocks for the different angles. The zero point may be tested by setting the instrument at zero and supporting the aligning board on some surface known to be level.

The inclination for the board used in the third method of aligning for lateral dihedral can be determined from the lateral dihedral table. Fifty inches make a convenient length for such a board in which case the Y (see Fig. 45) is just half of that given in the table for lateral dihedral angles.

Chapter 6

CARE AND INSPECTION

Cleanliness—Control cables and wires—Locking devices—Struts and sockets—Special inspection—Lubrication—Adjustments—Vetting or sighting by eye—Mishandling on the ground — Airplane shed or hangar — Estimating time — Weekly inspection card form.

CLEANLINESS.—One of the most important items is cleanliness of all parts of the plane. After every flight the machine should be thoroughly cleaned. To remove grease and oil from the wings and covered surfaces, use either gasoline, acetone or castile soap and water. If castile soap cannot be obtained, be sure the soap used contains no alkali or it will injure the dope. In using the gasoline or acetone, do not use too much or it will also take off the dope. A good way to use the latter is to soak a piece of waste or rag and rub over the grease or oil, then wipe off with a piece of dry waste. When using soap and water be careful not to get any inside the wing as it is liable to warp the ribs or rust the wires.

When mud is to be removed from the surfaces it should never be taken off while dry, but should be moistened with water and then removed.

Other parts of the machine should be kept thoroughly clean to keep down the friction.

Control cables and wires.—All cables and wires should be inspected by the rigger to see that they are at the correct tension. Also see that there are no kinks or broken strands in any of the cables or strands. Do not forget the aileron balance cable on top of the wings. When a wire is found to be slack do not tighten it at once but examine the opposing wire to see if it is too tight. If so the machine is probably not resting naturally. If the opposing wire is not over-tight then tighten the slack wire.

All cables and strands and external wires should be cleaned and re-oiled about every two weeks. The oil should be very thin so that it will penetrate between the strands.

Locking devices.—All threaded fastenings and pins should be inspected very frequently to see that there is no danger of anything coming loose.

Struts and sockets.—Since the struts are compression members, largely, they should be examined on the ends for crushing and in the middle for bending and cracking.

Special inspection.—A detailed inspection of all parts of the machine should be made once every week. Usually there is an inspection sheet provided for this purpose. If no sheet is obtainable, then one should be made before the inspection is started. Make a list of all the parts to be inspected, starting at a certain point on the machine, and following around until that point is reached again. When each part or detail is inspected it should be checked on the sheet as defective or O. K.

A good weekly inspection card form is given on pages 74 to 77.

Lubrication.—Always see that all moving parts are working freely before a flight is made. This includes undercarriage wheels, pulleys, control levers, hinges, etc.

Adjustments.—The angle of incidence, dihedral angle, stagger and position of the controlling surfaces should be checked as often as possible so that everything will be all right at all times. Alignment of the undercarriage should be made so that it will not be twisted and thus cut down the speed of the machine.

Vetting or sighting by eye.—This should be practiced at all times. When the machine is properly lined up, look at it and get a picture in your mind of just how it looks. Then when anything becomes out of line it can be easily detected without using any tools. See that the struts are in the same plane when looking at the front or side of the machine. The dihedral angle also can be checked by this method of sighting. Some flyers become so expert that they can check the alignment of the whole machine by eye.

Distortion

Always be on the lookout for dislocation of any of the parts. If any distortions cannot be corrected by adjustment of the wires, then the part should be replaced.

Mishandling on the ground.—Great care should always be taken not to overstress any part of the machine. Members are usually designed for a certain kind of stress and if any other kind is put upon them, some damage is likely to occur. When pulling an airplane along the ground, the rope should be fastened to the top of the undercarriage struts. If this cannot be done, then fasten the rope to the interplane struts as low down as possible.

Never lay covered parts down on the floor but stand them on their entering edges with some padding underneath. Struts should be stood on end where they cannot fall down.

Hangar.—The hangar at all times should be kept in the best possible condition. Never have oily waste or rags lying around on the floor or benches, as these are liable to catch fire. No smoking should be allowed in or near the building. Do not have oily saw dust spread around on the floor to catch the oil but have pans for this purpose.

In making replacements of defective parts, have a place for the old pieces. Never allow them to be put where they will be mistaken for new parts.

Each tool should be kept in a certain designated place and when anybody borrows a tool, be sure that he puts it back where it belongs.

Estimating time.—When any repairs are to be made, learn to estimate the time required for the job. With a little practice this can be done very accurately. It may help some time in making a report to an officer in charge as to when an airplane will be ready to go out again.

Weekly Airplane Inspection Card

Date....................191....

Airplane No. ....Make..................Model..............

Engine No. ....Make..................Model..............

Note: This card must be made out by Field Inspector for every machine under his charge, signed by him, and must be turned over to the Chief Inspector as soon as made out.

Landing gear:

Wire tension ....
Wire terminals ....
Strut sockets (nuts, bolts)............................
Loose spokes ....
Axles greased ....
Security of wheels to axle.............................
Shock absorber rubbers ....
Tire inflation ....

Propeller:

Condition of blades....................................
Hub assembly (bolts, washers, cotters).................
Security to shaft......................................
Thrust ....

Fuselage nose:

Tension fuselage bracing...............................
Tension and terminals wing drag bracing................
Engine bed and bolts...................................

Water system:

Leakage ....
Radiator full ....

Engine:

Valves—
Intake clearance ....
Exhaust clearance ....
Spark plugs—
Clean ....
Gap ....
Carburetor—
Security to manifold...................................
Bracing ....
Manifold joints ....

Oil system:

Leakage ....
Oil (grade) ....
Oil reservoir .... full

Magneto:

Mounting ....
Distributor board ....
Breaker point clearance.....................................
Transmission (drive) wear...................................

Throttle control:

Pulleys ....
Wiring ....
Bell cranks and connections.................................

Gasoline system:

Tank ....
Gasoline leads and connections..............................
Pump ....
Gasoline in tank....................................... full

Wing joints:

Lower wing—right ....
Lower wing—left ....
Upper wing—right ....
Upper wing—left ....

Wing wires: (tension, terminals, clevis pins, cotters, safety wires)

Flying wires—right wing ....
Flying wires—left wing ....
Landing wires—right wing ....
Landing wires—left wing ....
Wires, fittings, turnbuckles, cleaned and greased...........

Wing fittings: (bolts, nuts, cotters)

Right wing, upper.......................lower.................
Left wing, upper........................lower.................

Struts:

Sockets, bolts, cotters.....................................
Straightness ....
Right wing ....
Left wing ....

Ailerons:

Straightness ....
Hinge assembly (lubricate with graphite grease).
Security ....
Wear ....
Hinge pins and cotters......................................
Control wire connection (mast)..............................
Frayed control wire (wheel).................................
(pulleys and guides)..

Note: Control wires frayed at any part of their length must be replaced at once.

Pulleys ....

Greased ....

Free running ....

Right ailerons, upper....................lower....................

Left ailerons, upper....................lower....................

Fuselage rear interior:

Wire tensions ....

Longerons ....

Fittings ....

Alignment ....

Stabilizer:

Bolts, nuts, cotters, braces..

Vertical fin:

Bolts, nuts, cotters, braces..

Rudder:

Hinge assembly ....

Security ....

Wear ....

Hinge pins and cotters..

Control wire connections..

Mast ....

Footbar ....

Frayed control wire..

Note: Control wires frayed at any point of their length must be replaced at once.

Pulleys ....

Greased ....

Free running ....

Elevators:

Hinge assembly ....

Security ....

Wear ....

Hinge pins and cotters..

Control wire connections—Mast..

Control wire connections—post..

Frayed control wire..

Note: Control wires frayed at any point of their length must be replaced at once.

Pulleys ....

Greased ....

Free running ....

Right elevator ....

Left elevator ....

Tail skid:

Skid ....

Fittings ....

Shock absorber ....

Controls:

Free and proper operation (lubricate with graphite grease)...........

Elevator ....

Rudder ....

Aileron ....

Alignment of entire machine:

....

....

....

....

(Signed) Field Inspector.

Chapter 7

MINOR REPAIRS

Patching holes in wings—Doping patches—Terminal loops in solid wire—Terminal splices in strand or cable—Soldering and related processes—Soft soldering—Hard soldering—Brazing—Sweating—General procedure in soldering—Fluxes—Melting points of solders.

THE materials used in patching holes in linen-covered surfaces is unbleached Irish linen, the same kind as used in covering the wings. The material must be unbleached or it will not shrink the required amount. Generally the kind of dope used is Emaillite dope, although the acetate or nitrate dopes could be used. The dope should be applied in a very dry atmosphere or on a sunshiny day at a temperature not less than 65 deg. F. A brush or a piece of waste may be used to apply the dope.

In patching a hole the first thing to be done is to clean the surface of the old dope. To do this, fine sand paper may be used or acetone, gasoline or dope. In using the sand paper, care should be taken not to injure the covering. When using the acetone or gasoline, it should be put on the surface, allowed to stand for a while to soak up the old dope, then scraped off. The same method is applied when using dope to clean the surface.

After the surface is cleaned, the edges of the hole should be sewed if it is of any considerable size. To do this sewing linen thread and a curved needle are used. The stitches should not be closer together than ½ in. and far enough back from the edge so that there is no danger of their tearing out. With a small hole, such as a bullet hole for instance, it is not necessary to do any sewing. When the hole is several inches square, a piece of unbleached linen should be sewed in to give a body for the top patch so that it will not be hollow in the center after it is dry. The sewing up of holes should be done after the surface is cleaned so that any slackness may be taken up before the patch is applied.

After sewing is finished the patch is cut. It should be made about 1 to 2 in. larger on each side than the hole. The edges of the patch must be frayed for about ¼ in., this being done to prevent them from tearing easily.

Dope should now be applied to the wing. Generally several coats are put on so that there will be a sufficient amount to make the patch stick well. After the last coat is applied the patch should be put in place immediately before the dope has a chance to dry. Any air bubbles and wrinkles should now be worked from under the patch by rubbing with the fingers, and more dope put on top of the patch. Usually there are six or seven coats of dope applied on top of the patch, allowing time for each coat to dry before another is applied.

Any small amount of slackness in the patch will probably be taken out as the linen shrinks. If the patch is hollow after the dope is thoroughly dry, however, it is not a good patch and should be removed. A good patch should be tight around the edges as well as in the center over the hole and should contain no creases or air bubbles.

Terminal Splices

A loop or splice must be formed in the end of every brace wire or control cable where it is attached to a strut socket, turnbuckle, control mast, or other form of terminal attachment. The manner of making the loop or splice in the wire will vary according to the type of wire or cable used. The terminal in the end of a solid wire is made in the manner shown in Fig. 33.

There are several points to be observed in making this type of terminal splice, as follows: (a) The size of the loop should be as small as possible within reason, as a large loop tends to elongate, thus spoiling the adjustment of the wires. On the other hand, the loop should not be so small as to cause danger of the wire breaking, due to too sharp a bend. (b) The inner diameter of the loop should be about three times the diameter of the wire, and the reverse curve at the shoulders of the loop should be of the same radius as the loop itself. The shape of the loop should be symmetrical. If the shoulders are made to the proper radius there will be no danger of the ferrule slipping up towards the loop. (c) When the loop is finished it should not be damaged anywhere. If made with pliers there will be a likelihood of scratching or scoring the wire, which would weaken it greatly. Any break or score in the surface coating of a wire destroys the protective covering at that particular point and the wire will soon be weakened by exposure. A deep nick or score would greatly weaken the wire and eventually result in breakage at that point.

Splicing a strand or cable.—The splice in the end of a strand or cable is entirely different from the terminal of a solid wire. The end of the strand is led around a thimble and the free end spliced into the body of the strand or cable just below the point of the thimble. Such a splice is afterward served with twine, but the serving should not be done until the splice has been inspected by whoever is in charge of the workshop. The serving might cover bad workmanship in the splice.

Soldering.—Terminal loops or splices in solid wire and also splices in the ends of strand or cord are sometimes soldered after being formed. There are some objections to soldering at these points, however, as outlined on page 58. The ensuing instructions for soldering work will prove valuable in cases where this method of securing a terminal splice is considered desirable.

Joining of metals by soldering and related processes.—There are several methods of joining metals together by alloys which melt at a lower temperature than the metals to be joined. These processes differ in the alloys used and in their melting temperatures. They are divided into four classes, as follows:

(1) Soft soldering.—This method is the one used in tin-smithing generally, where the solder is melted by means of a hot soldering copper over the surfaces to be joined. The solder used in this process has a low melting point.

(2) Hard soldering.—This method is usually used in jewelry work and in the arts, where a higher temperature must be withstood. The joining metal in this case has a much higher melting point than soft solder, and must be heated with a blow-torch to make it flow.

(3) Brazing.—This process differs from hard soldering only in the fact that the joining metal has a still higher melting point. It is used principally in motorcycle, bicycle and automobile construction, where greater strength is required.

(4) Sweating.—This is a process used where the pieces to be joined can first be fitted together, then individually coated with solder, then clamped together and heated until the solder flows and cements them solidly together. This method allows for a more perfect joint being made. The more accurately the parts are fitted together the stronger the union will be. Also, the thinner the coat of soldering material, within reasonable limits, the stronger the joint.

All of the above methods are used more or less in airplane construction and maintenance, but the one that is most generally used is the first, or soft-soldering method.

Cleanliness is of prime importance in making joints or fastening by any of these methods. In soldering, the first step is to see that the soldering copper is clean and well tinned, for this may determine the success or failure of the job. There are several ways of cleaning and tinning the soldering copper, but the one recommended is to heat the copper to about 600 deg. F., then dip the point quickly into a cup or jar containing ammonium chloride (NH_4Cl) and granular tin or small pieces of solder. If any considerable amount of work is to be done, an earthen jar or a teacup can be used, and kept partly filled with this mixture.

Tinning Soldering Coppers

Another way of tinning the soldering copper is to make a depression in a piece of sheet tin and place in it a small quantity of soldering flux together with a piece of solder. File the copper until bright, heat it to about 600 deg. F., and then move it around, while hot, in the depression in the tin until it becomes coated with molten solder. It will now be ready to use.

The next step is to clean thoroughly the parts to be joined, using fine emery cloth, sandpaper or a scraper. If the parts are of raw material, sandpaper will do, but if they are old parts which previously have been exposed, or if a heavy oxide has formed, the surfaces to be soldered should be filed or scraped until perfectly bright and clean. The cleaned surface should then be covered with soldering fluid or one of the many soldering pastes.

Heat the soldering copper to about 600 deg. F., and touch it to the solder, being careful to get only a small amount of solder on the copper. Rub the copper over the surfaces to be joined until a bright, even coating of solder clings to the surfaces. Place the pieces together and heat until the solder flows, using the hot copper to furnish the necessary heat and adding more solder as needed. Care must be taken not to overheat the pieces at the joint, as this has a tendency to weaken the metal at that point and may cause trouble.

The same general procedure as the above is followed for hard soldering, with the exception that a higher temperature must be applied.

Fluxes

Fluxes are used in soldering to prevent, so far as possible, the formation of oxides on the heated surfaces, and to flux off those that may have formed. Acid fluxes are the most effective and on iron or steel are practically necessary. The objection to their use is that unless the parts are thoroughly cleaned after soldering the acid in the flux attacks and corrodes them.

In the case of stranded wires or cables the flux will penetrate into the minute spaces between the strands and will be extremely difficult to remove or neutralize, even when the cable or wire is washed with or dipped in an alkaline solution, such as soap or soda water.

Some of the fluxes in general use are:

Zinc chloride (Zn Cl), corrosive.

Dilute muriatic acid (H Cl), corrosive.

Resin, non-corrosive. This is satisfactory for tin, but will not work on galvanizing.

Resin and sperm candle melted together make a fair non-corrosive paste. For either tin or galvanizing use three parts resin to one part sperm candle. Sometimes better results are obtained on dirty surfaces by adding one part alcohol to this mixture.

Melting points of solders.—The melting points of solders composed of tin and lead in various proportions are as follows:

Proportion		Melting Point
Tin	Lead	
1 part	25 parts	558 deg. F.
1 part	5 parts	511 deg. F.
1 part	1 part	370 deg. F.
1½ parts	1 part	340 deg. F.
5 parts	1 part	278 deg. F.

A composition of 1 to 1 is most commonly used for tin-smithing. For electrical work where the solder is used in the form of wire, a proportion of 1½ to 1 or 2 to 1 is used.

Chapter 8

INSTRUMENTS

Compass—Magnetic errors—Variation—Deviation—Corrections—Napier diagram—Heeling errors—Magnetic clouds—Aneroid barometer—Errors—Uses—Altimeter—Errors—Banking meter—Air speed indicator—Incidence indicator—Sperry Clinometer—Automatic drift set—Bourdon gauge—Radiator thermometer.

UNDOUBTEDLY the compass is one of the most important instruments on an airplane and a thorough knowledge of its construction, operation and theory are absolutely essential to enable the student to become a successful pilot.

Theoretically, a compass is a permanently magnetized needle swinging freely in a horizontal plane on a pivoted point and indicating a north and south direction, due to the attraction of the earth's magnetic north pole for the north pole of the compass needle, and the repulsion of the magnetic pole for the south pole of the compass needle. As is the case with all mechanisms, this theory has been combined with modern refinements, until we find in practice at the present time two main types of compass, viz., the dry card compass and the liquid compass.

The dry card compass consists of a graduated card over which the magnetized needle swings freely. The accompanying cut (Fig. 47) illustrates such a graduated card, which, in a much smaller form, is in use as a pocket compass. It can readily be seen that such an instrument is very easily disturbed, the slightest jar starting it dancing. Gun fire would, of course, render it useless. Another very prominent defect is its inability to come to a complete rest without considerable oscillation. To overcome these mechanical errors the U. S. Navy produced the liquid compass now used on all vessels, and, in a more perfected form, on airplanes.

The construction is somewhat more complex as illustrated by Fig. 46, which is a cross-section of a Creagh Osborne standard airplane compass. Instead of a needle swinging over a graduated card, this compass has a graduated card swinging past a certain point or line marked on the inside of the compass case, called the "lubber line," the entire mechanism being immersed in an airtight bowl of liquid.

In detail the construction is as follows: M is a pivot supporting the graduated dial P, the latter being made of tinned brass and graduated in quarter points and half degrees. This dial is hollow and airtight, having in the center a spheroidal air chamber Q to buoy up the weight of the card and magnets O, allowing a pressure of between 60 and 90 grains on the pivot at 60 deg. F. The pivot rests in a socket whose bottom N is a sapphire jewel to form a bearing. At the base of the column is a telescopic spring T to ease the jars on this jewel. The magnets O consist of four highly magnetized bundles of steel wires contained in a sealed cylindrical case, the magnets placed parallel to the north and south line of the card, and so aligned that their entire directive force will cause the north point of the card to point to magnetic north. The card is mounted in a cast bronze bowl D which is entirely filled with liquid—45 percent alcohol and 55 percent distilled water. Beneath the bowl is a self-adjusting expansion chamber K of elastic metal, arranged with a small hole L, to permit circulation of the liquid between the bowl and expansion chamber. The expansion chamber is so designed as to expand with a rise in temperature and counteract the expansion of the liquid in the bowl, thus keeping the bowl free from bubbles. At one side is a filling screw through which the air bubbles, which are bound to appear, may be removed, and additional liquid added.

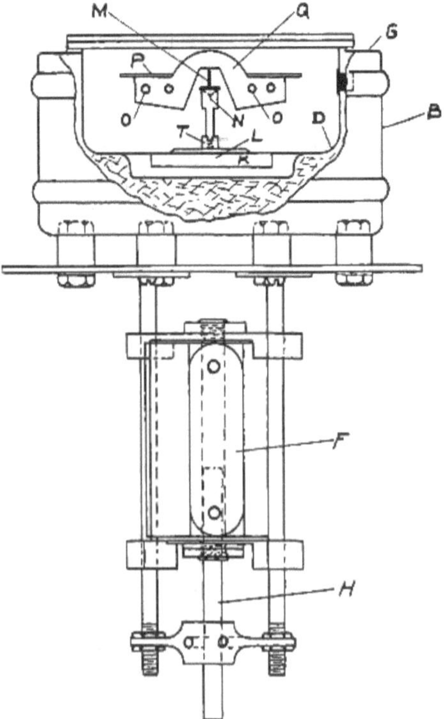

Fig. 46—Details of the Creagh Osborne airplane compass

The graduations on the card are marked every 20 degrees, the 0 being dropped to avoid crowding. These markings are insoluble in the liquid and are visible at night. The compass bowl sets in an outer bowl B of the same material. It rests on rubber supports, and between the two bowls is a packing of horsehair for absorbing shock. Beneath the compass are the corrections for various magnetic effects which will be discussed later.

It will be recalled that the dry card compass was subject to oscillation and was easily disturbed. It is obvious that in the liquid compass both these errors are materially reduced, the liquid tending to dampen

Fig. 47—Standard compass card

the oscillatory motion, or magnetic moment, as it is called, and also preventing the needle from dancing about the pivot point. In addition, the combined effect of the small but powerfully magnetized wires is to give the system a greater sensibility or directive force.

Magnetic Errors

Variation.—The errors previously discussed have been mechanical in nature and within the compass itself. There is another class of errors, affecting compasses of all types, which are due to external influences, and a thorough knowledge of their various actions is absolutely essential to the use of the instrument.

These may be called "magnetic" errors, inasmuch as they are due either directly or indirectly to the magnetic influence of the earth itself.

Consider the earth as a huge magnet having a north and south magnetic pole. The north magnetic pole and the north geographical pole do not coincide in location. The true north or geographic north is what we call the north pole and is the northernmost point on the earth's surface. The magnetic north pole is located near the northern part of the Hudson Bay region, and is so called because the earth's magnetic influences in the northern hemisphere reach a maximum at this point. Inasmuch as the north-seeking end of the compass needle points to this point, it can readily be seen that there is an error between the true north and the compass reading. This error is called the variation and is known and tabulated for all points on the earth's surface. It changes slightly from year to year, but for air work we can consider it as constant for any point, and, as the compass cannot be mechanically corrected for it, the variation must always be included in the calculation of bearings.

Depends on Locality

Variation may be either easterly or westerly, depending on the location on the earth's surface; if easterly, it is given a positive sign, if westerly it is considered negative in sign. These signs or directions are determined by the observer who is considered as being at the center of the compass and facing the point under consideration. It must of course be understood that all conditions are reversed in the southern hemisphere. In this connection it might be interesting to note that the variation in Flanders amounts to about 14 deg. and is westerly. Hence to fly true north from any point in that region the compass will read 14 deg. [360 deg. – (–14 deg.)]. The compass being graduated from zero to 360 deg., the following rule may be formulated for finding the compass course from the true course: "Always subtract the variation (neglecting sign) if easterly, and add it, if westerly, from the true course, in order to get the compass course."

Deviation.—Again let us consider the earth as a huge magnet with lines of force flowing between the poles like invisible rubber bands. If an iron rod is held parallel to these lines of force, i. e., approximately north and south, and then struck two or three sharp blows, it will be found by bringing it to a compass needle, that one end either attracts or repels the needle, thereby indicating that the rod has become magnetized. Now, if the rod be reversed, end for end, and the experiment repeated, it will be found that the magnetism has become reversed. In other words, there has been induced into an iron rod, a certain amount of magnetism, which is called "induced magnetism." If the rod is of hard iron it will remain a permanent magnet, but if it is of soft iron it will lose the magnetism as soon as the inducing influence is removed. It readily can be seen from the above experiment that in the construction of an airplane there is a certain amount of magnetism induced into the iron and steel parts of the engine, beams, wires, etc., due to hammering. Such a magnetism is known as "residual" or "sub-permanent" magnetism, and the error which it produces in the compass is called "deviation."

Correction for sub-permanent magnetism.—There are two methods of correction for sub-permanent magnetism, both of which are rather technical in theory, but simple in practice. The first is by means of small "adjusting magnets," and the second is "Napier's diagram." No attempt will be made to explain the theory of either method, but simply how they operate in actual practice. It should be added that Napier's

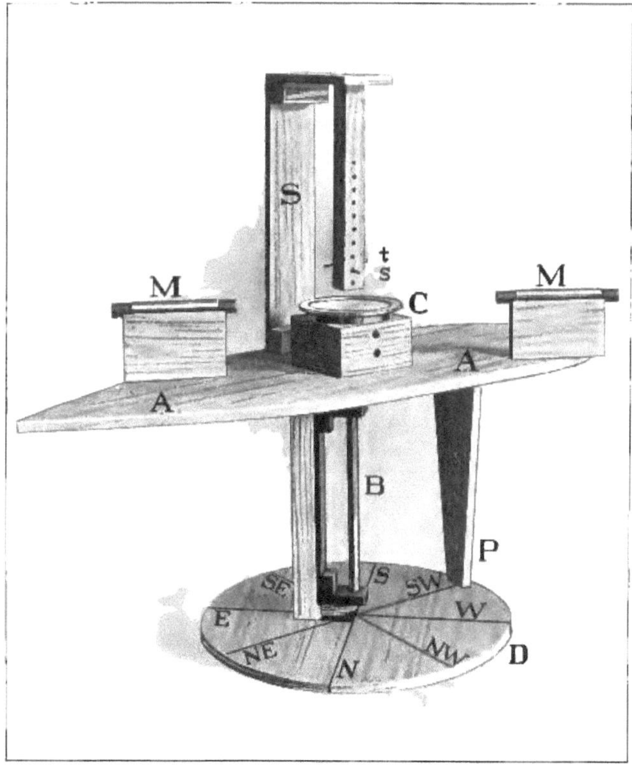

Fig. 48—Model showing how corrections are made for residual or sub-permanent magnetism

diagram is a simple method of finding a table of corrections rather than elimination of the deviation.

Fig. 48 is a model which serves to illustrate the method of correction for sub-permanent magnetism by means of small adjusting magnets. AA represents a ship which can be rotated in a horizontal plane above the axis B and so placed that the center line of the ship has the true magnetic bearing indicated by the dial D. The compass is fastened to the block which rotates with the ship AA. The sub-permanent magnetism is represented by the magnets MM which can be placed in any position. Practice has shown, however, that one magnet is better than two. These magnets are used merely to illustrate the method. In actual practice the metal parts of the ship or airplane take their place.

Small correcting magnets s and t are held by the standard S over the vertical axis of the compass. Their distance can be varied by means

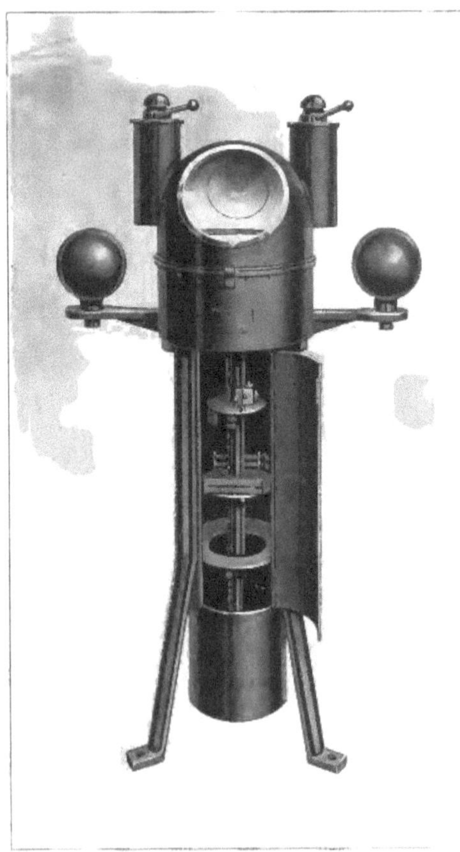

Fig. 49—Mariner's compass

of the small holes as shown. These small magnets may be placed either above or below the compass as long as they are in the plane of its vertical axis. The compass is then corrected by raising or lowering these magnets. The following directions explain in detail how an airplane compass is corrected:

1. Place the airplane in the true magnetic north and south flying position (almost every airdrome has some cement cross or stakes to mark these positions).

2. Bring down one of the small adjusting magnets over the compass in an east or west position. If deviation increases reverse the magnet end for end and again lower it over the compass until the deviation is zero.

3. Fix the magnet at that height above the compass at which deviation is zero.

4. Turn the plane to a true magnetic east and west flying position.

5. Repeat 2 and 3, using another adjusting magnet and being sure to apply it in an east and west position.

6. Check on other magnetic bearings.

In summarizing this method there are a few salient facts to be emphasized. The adjusting magnets must always be applied in an east and west position regardless of the heading of the ship or plane. After the position of a correcting magnet has been determined it can be turned in any direction simultaneously with the compass without destroying its effect. In actual practice the corrections are usually placed below the compass in receptacles arranged for that purpose. This is shown in Fig. 49, which is an illustration of a mariner's compass, or by referring back to Fig. 46, F. The correction is not a permanent one, and from time to time should be checked.

Napier Diagram

Napier diagram.—The second method of correcting for sub-permanent magnetism is by means of Napier's diagram, an illustration of which is shown on page 87. The diagram is nothing more or less than a graduated compass card straightened out into a straight line. This method does away with the small iron correctors. Referring back to Fig. 48, assume an airplane AA rotating in a horizontal plane. The magnets MM represent the sub-permanent magnetism causing deviation of the compass; in practice the metal parts of the engine and plane have the same effect.

To plot deviations on the Napier diagram.—Obtain the compass reading for the true magnetic course at each of the cardinal points of the compass. If the reading is greater than the true course the deviation should be plotted on the left side of the chart. If the reading is less than the true course, plot the deviation on the right side of the chart. Mark on the center scale of the diagram the compass reading for a given true magnetic course, north or 0 for instance, then draw a line, paralleling the dotted lines, from the compass reading to the solid line from the corresponding true magnetic course. Follow the same procedure for each of the cardinal points of the compass for which readings were taken, then draw a curve through the points at which the dotted lines projected from the compass readings intersect the solid lines from the corresponding true magnetic courses. From this curve the deviation for any course can be plotted.

To obtain course to be steered for a given true magnetic course.—From the point on the center scale representing the true magnetic course desired project a line parallel to the solid lines until it intersects the curve already plotted. From this point project another line parallel to the dotted lines back to the center scale. The point at which this last line intersects the center scale gives the course to be steered.

A homely, but easily remembered rule for the use of a Napier diagram is given below:

Ship's Head		Compass-Degrees	Deviation-Degrees
Magnetic	Degrees		
N	0	28	-28
N.E.	45	80	-35
E.	90	118	-28
S.E	135	145	-10
S.	180	171	+9
S.W.	225	199	+26
W.	270	236	+34
N.W.	315	289	+26
N.	360	388	-28

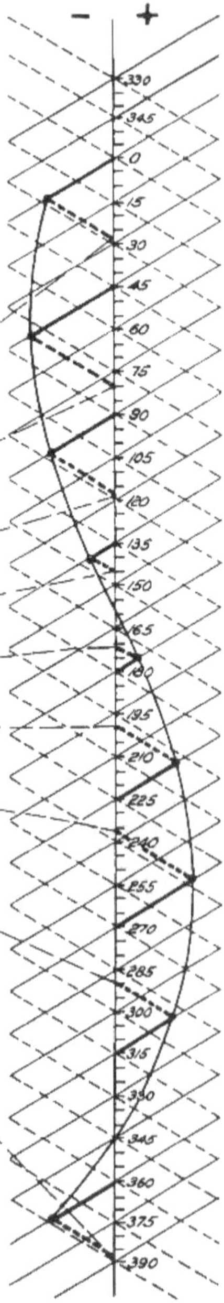

"If you wish to steer the course allotted,
Depart by plain, return by dotted;
From compass course, magnetic to gain,
Depart by dotted, return by plain."

Fig. 30—Napier diagram showing method of plotting deviation curve

A Napier diagram is shown in Fig. 50 together with a sample table giving assumed compass readings for the cardinal compass points, and showing how the curve is plotted from these compass readings.

Heeling error.—In all previous discussions the force of sub-permanent magnetism has been considered as acting only in a horizontal plane. It can readily be seen from the accompanying sketch that, when the vessel heels to port or starboard the horizontal and vertical components

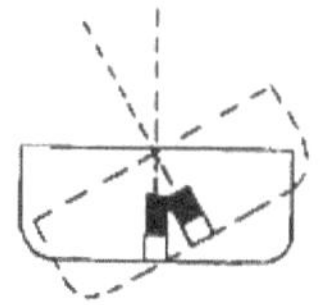

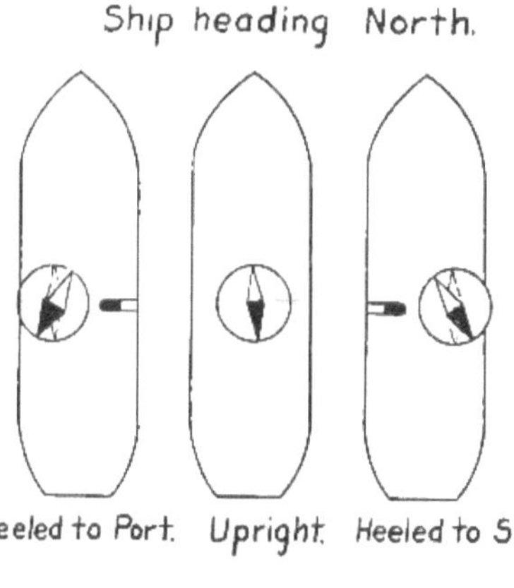

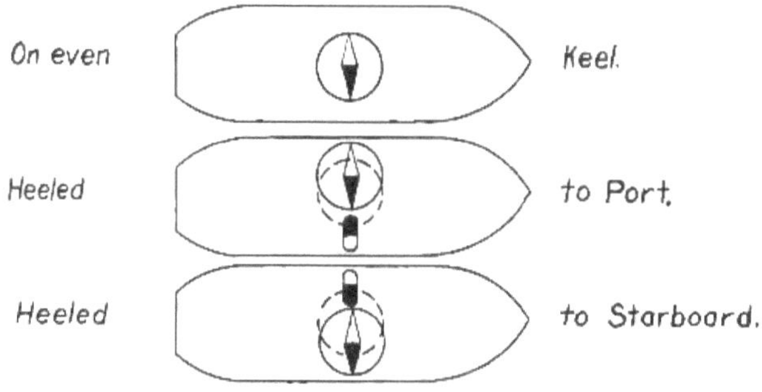

Fig. 51—Showing compass error due to heeling of ship when headed north or south

change position to a certain extent, producing a resultant force which acts on the needle. This is the "heeling" error and as the illustration shows, it is a maximum when the ship is heading north or south and zero when east or west. Being due to vertical forces it is corrected by a single vertical magnet placed directly in the vertical axis of and at a certain distance below the compass while the airplane is upright. This is illustrated by H in Fig. 46. In the northern hemisphere the north pole of this magnet will be up.

Magnetic clouds.—A third form of error, more or less mechanical, should also be mentioned. It is based on the erroneous supposition that the clouds exert a magnetic influence on the compass needle, due to the fact that on entering a cloud, the card may be observed to turn of its own accord. The theory that clouds exert a magnetic influence has been put forth in explanation. Ordinary clouds are not magnetic — only thunder clouds—and the aviator does not usually fly in thunderstorms.

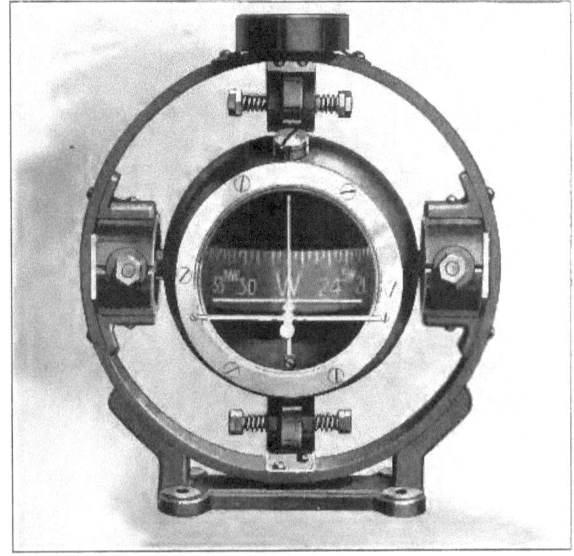

Fig. 52—Standard form of airplane compass

The magnetism of the earth causes the north end of the needle or card to dip in the northern hemisphere, the angle varying from 0 deg. at the equator to 90 deg. at the magnetic north pole. To offset this, the south pole of the compass is weighted slightly. If a machine be flying north, and turns to right or left, centrifugal force acts on the needle, causing the southern end to swing out and the northern end to swing in, when exactly the reverse should take place. If a pilot is unaware of this action, he is very liable to increase his turn still more, thereby increasing his error. In time the momentum may cause the card to swing completely around, as is often the case. If the aviator has any point to steer by he can hold the machine steady until the card has settled down. The only way to correct this error is to make the needle a very weak magnet, thus requiring a very small weight on the south end to offset the dip.

The aviator should examine his compass with a view to finding out if it is subject to this error. Compasses with a short period of vibration are likely to be wrong; those with a long period are usually right.

Summary.—This completes the discussion of the main errors of the compass. There remain a few forms of error which are to be studied only in the case of the mariner's compass, and which are too delicate to be considered in aerial work. The student should acquire the habit of constantly checking the instrument when in flight by bearings or rivers, roads, bridges, towns, etc., over which he passes. This cannot be too strongly emphasized.

Figs. 52 and 53 are illustrations of two forms of compass now in use on airplanes, the latter being a Creagh Osborne type, a cross section of which is shown in Fig. 46.

The Aneroid Barometer

Fundamental principles of air pressure.—If a glass tube sealed at one end and completely filled with mercury is inverted into a bowl of the same liquid and allowed to run out, it will be found that the mercury in the

Fig. 53—Creagh Osborne type of airplane compass

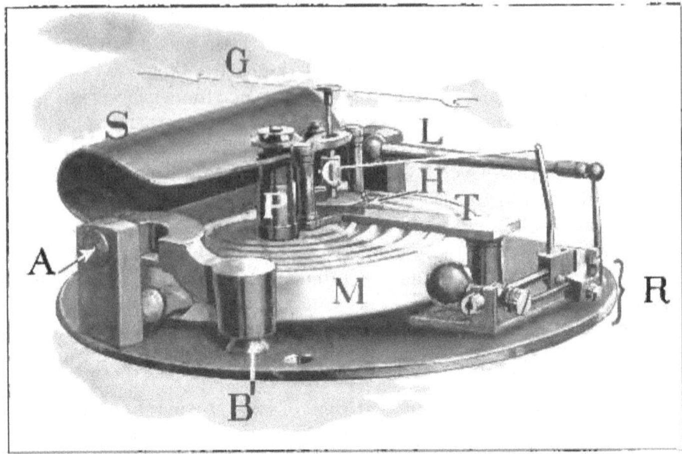

Fig. 54—Aneroid barometer with cover removed

tube drops a total of approximately 30 in. This experiment indicates that the air has weight, this weight exerting a pressure sufficient to maintain the column of mercury in the tube up to about 30 in. from the top. Calculation shows that the pressure amounts to about 14.7 lbs. per square inch at sea level. This experiment also illustrates in a simple way the theory and construction of the mercurial barometer, which is used in all measurements of atmospheric pressure. At the same time it readily can be seen that such an instrument would be entirely impracticable for use on an airplane. Hence for use in aerial work there is used what is known as the aneroid barometer, an illustration of whicl is shown in Fig. 54.

Construction of aneroid barometer.—The aneroid barometer receives its name from the word "aneroid" which means "without fluid." Its construction is very simple, and in detail is as follows: In place of the column of mercury there is a German silver corrugated expansion chamber M (see Fig. 54) connected by a post P to a steel leaf spring S which in turn is connected to a pivot A. The spring is also connected by a lever arm L and a system of levers R to a post supported by a table T. This has a chain gear mechanism at C operating the indicator G. A hair spring H is fitted which serves to take up any slight shock due to loose connections. Any elevation or depression of the expansion chamber due to change in barometric pressure raises or lowers the leaf spring, which in turn communicates this action by means of the lever arm and system of levers to the indicator. The leaf spring also serves the purpose of maintaining a constant tension upon the expansion chamber and the indicator.

Errors.—The aneroid is not an absolutely accurate measurement of air pressure. It does not permit of the numerous refinements of mechanism necessary to compensate for the various conditions that must be taken into consideration in atmospheric measurements. It is absolutely necessary that the entire system be rigidly connected, otherwise any small change in the expansion chamber would merely serve to overcome the slack in the mechanism. In other words there must be a perfect balance of the system. For this reason sudden changes in temperature affect it very seriously, bringing about looseness in the joints, and increases in friction. As it is not designed to compensate for these changes, the readings may often be in error. By tapping the aneroid at different points, errors due to slack or lack of balance, may be detected.

Uses of the aneroid barometer.—The most common use of the aneroid in aviation is for determining changes in elevation, in which use it generally appears as a pocket aneroid. It is very frequently used in connection with sounding balloons for determining atmospheric conditions at higher levels. Such aneroids are self-recording and are known as barographs.

The Altimeter

The theory, construction, and operation of the altimeter are practically the same as the aneroid barometer. It is primarily an instrument for measurement of altitude and is very often combined with the aneroid in which case it is known as the aneroid altimeter. There are, however, certain conditions that must be considered in the graduations of the altimeter that are not considered with the aneroid. It was shown that the atmosphere exerts a pressure of approximately 14.7 lbs. per square inch at sea level. Observation has proven that this pressure does not vary directly with the altitude. In fact, half the earth's atmosphere is included in the first three and one-half miles above the earth's surface, and it is assumed that traces of our atmosphere can be found even as high as two hundred miles. From this it readily can be seen that at an altitude of two miles the pressure will not be one-half as much as at one mile. The pocket aneroid altimeter is accordingly graduated to meet these conditions, smaller graduations occurring at the higher altitudes.

The altimeter in general use in aerial work is graduated as in Fig. 55. Note that the graduations are equal. This is compensated for by a somewhat complicated system of levers within the instrument itself. The altimeter is so constructed that it may be adjusted for any elevation above sea-level at which the pilot may be, this being done by rotating the dial until the zero and needle are coincident. This adjustment is provided to enable the pilot to tell at a glance his actual altitude above the ground, without having to make allowances for the height of the ground above sea-level. This latter might be a very few feet or several thousand, depending on the locality.

Errors of the altimeter.—The altimeter is subject to the same errors as the aneroid but it is highly important that these errors be taken into consideration far more carefully than in the latter instrument. Temperature changes will affect the looseness in the links thereby destroying the balance of the instrument. Sudden changes are even more injurious.

Changes in barometric pressure after the pilot has started must be taken into consideration and no altimeter yet has been designed which will do this. If during the course of a flight the aviator should pass from a region of high barometric pressure to one of low pressure without changing his altitude, he would find that his instrument was registering an increased height. This is due, of course, to atmospheric conditions, of which the aviator must be a student in order to be on guard for such errors. Weather maps for the region in which he is operating will give him some idea of the location of regions of high or low barometric pressure.

Another very important error over which the altimeter has no control originates from the topography of the ground over which the machine

Fig. 55—Altimeter as used on airplanes

is flying. For the sake of illustration we will assume that an aviator leaves Omaha, which has an elevation of 3000 ft. above sea-level, and flies to Pike's Peak. He will set his instrument at zero at Omaha, and on alighting on Pike's Peak, which is about 14,000 ft. above sea-level, will find that his altimeter reads in the neighborhood of 11,000 ft. That is, he is 11,000 ft. above Omaha but not above Pike's Peak. In other words, the altimeter does not take into consideration changes of elevation in the territory over which a plane flies. From this it can be seen that a thorough knowledge of the topography of the region in which the pilot is operating is very essential in aerial work.

There is a fourth form of error which is mechanical and is due to the lag of the instrument. It occurs especially in connection with the nose dive, and may be the cause of a very serious accident. The aviator in dropping from an altitude of 10,000 ft. to 1,000 ft., for instance, will reach the lower elevation before his instrument can record the change. The danger lies in the fact that if he depends entirely upon his altimeter, he may not realize until too late that he is much closer to the ground than the altimeter indicated. This error will probably never be entirely eliminated, but it gradually is being reduced with increasing perfection in construction.

This practically covers the main errors of the instrument, all of which are due to external influences. Mechanical errors can be detected either by comparing the instrument with a standard barometer or by tapping it to discover slack. In fact, the instrument frequently should be compared with a standard mercurial barometer, merely as a check on its accuracy.

It should be added that since hot air weighs less than cold air, the readings of an altimeter for the same height will vary with the temperature of the air. Obviously it is impossible to obtain this temperature at all times, hence this factor must be neglected, thus introducing an error which may reach as high as ten per cent.

Banking Meter

As its name indicates the banking meter, or banking indicator as it is sometimes called, is used to show the amount of bank which the pilot is making and guard him against skidding or side-slipping.

There are two forms in use at the present time. One form of this meter is constructed on the same principle as the ordinary spirit level such as is in use on carpenter's levels, transits, etc. It consists essentially of a small glass tube nearly filled with liquid, leaving a small bubble. When the tube and its mounting are horizontal the bubble will come to rest in the center of the length of the tube. Any slight inequality in the surface being tested is sufficient to send the bubble to the end of

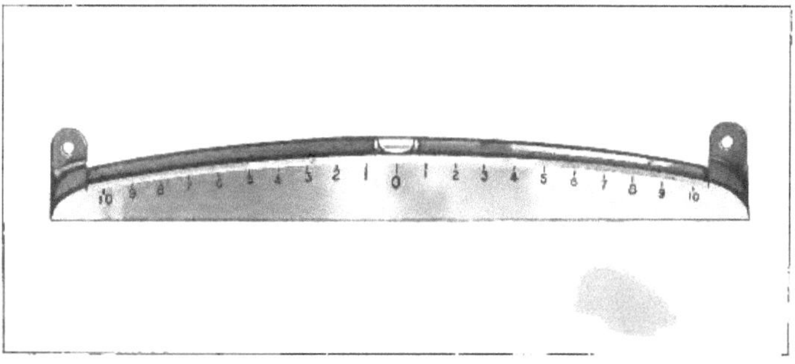

Fig. 56—A form of banking meter patterned after the spirit level

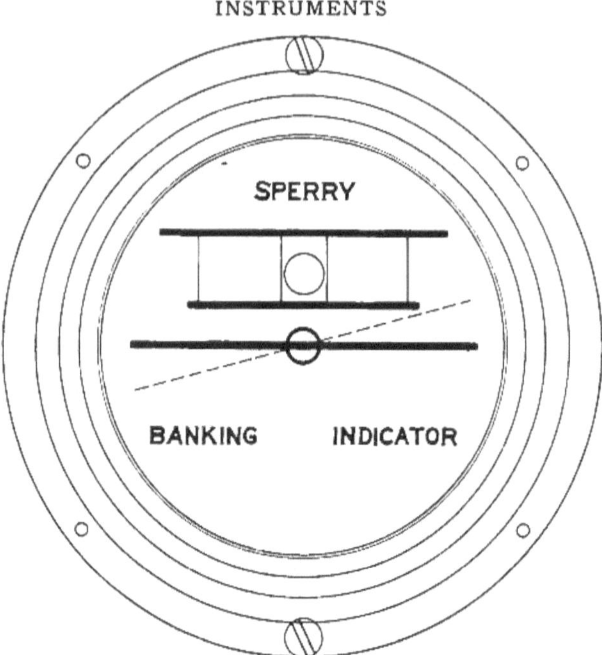

Fig. 57—The needle or bar, pivoted in the center of this banking indicator, is actuated by a plumb bob as the plane banks

the tube. Such an instrument would be impracticable in an airplane, hence a curved tube is used. The fault of this instrument lies in the fact that it does not indicate the amount of bank correctly as a rule.

The Sperry Gyroscope Co. has perfected a banking indicator, an illustration of which is shown in Fig. 57. It is based on the principle of the plumb bob. The upper framework shown on the dial indicates the position of the airplane relative to the horizontal and the lower indicator (the horizontal bar pivoted at its center) tells the pilot whether he has banked his machine sufficiently to meet the centrifugal force due to a turn. The dials are painted with radiolite and are visible at night.

The inclinometer, which may be used to indicate the longitudinal or lateral inclination of the plane, is practically the same in principle as the first form of banking meter described and merely serves as a general check on the attitude of the machine in flight. Instruments based on the principle of the spirit level are very inaccurate, being subject to error due to acceleration. Fig. 56 is an illustration of the airplane inclinometer. (See also incidence indicator and elinometer.)

Air Speed Indicator

The construction of this instrument is based on the theory of a speed measuring device known, after its inventor, as a Pitot tube. Such an instrument contains two essential elements: the first is the dynamic opening or mouth of the impact tube. (See Fig. 58.) It points directly against the current of gas or liquid in which the speed is to be measured and receives the full impact of the current. The second is a static opening for obtaining the so-called static pressure of the moving fluid, that is, the pressure that would be indicated by a pressure gauge moving with the current and not subject to impact. To avoid the influence of impact the static opening is placed at right angles to the dynamic opening. If the two pressure heads are connected to the opening of a U tube partly filled with liquid, and a current of air generated against the mouth of the impact tube it will be seen that the liquid will indicate a difference in pressure between the two openings. Starting with the formula $V = \sqrt{2gh}$ where h is the difference in elevation between the levels, and g the measure of acceleration due to gravity, a measurement can be obtained of the velocity of the air current by measuring the difference in pressure between the two openings. Fig. 58 is a cross section of the opening in a Pitot tube. While the dynamic pressure can be accurately determined, the static pressure is very uncertain, as the air rushing by the static opening is apt to cause suction. The instrument should therefore be calibrated.

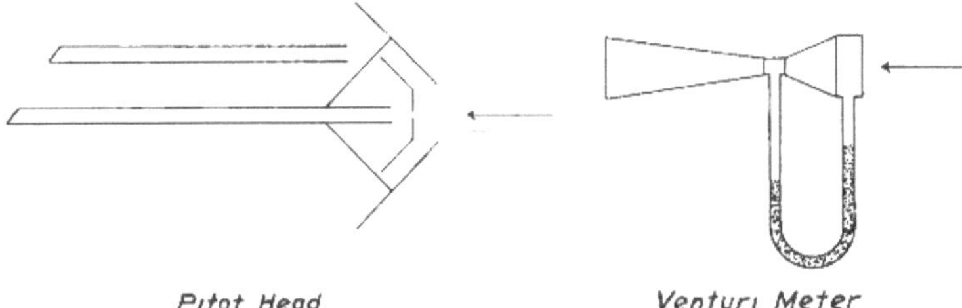

Fig. 58—Showing operating principles of two forms of air speed meters

Another form of air speedometer, also shown in Fig. 58, is based on the principle of a venturi tube. This tube consists of a short converging inlet followed by a long diverging cone. The opening is placed at right angles to the air current and a measurement is made of the difference in pressure existing between the opening diameter and the smallest diameter of the throat. This measurement is based on the ratio of entrance to throat area, these being the names of the opening and of the smallest area. The tube is provided with connections to a differential gauge by which this difference in pressure is measured.

The Incidence Indicator

The incidence indicator, as its name suggests, is used to measure the angle of attack. For this purpose an instrument, if possible, should be "dead beat," or in other words, free from any tendency to swing, and it must actually register any change in the direction of the flow of air to the supporting wings or surfaces.

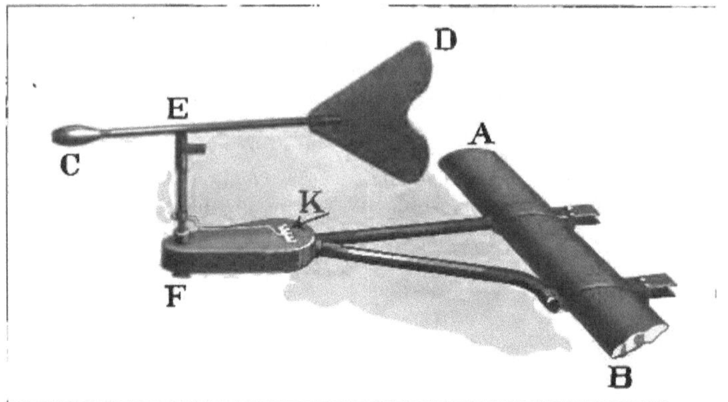

Fig. 59—Incidence indicator mounted on one of the forward struts. It operates colored lights in the dial shown in Fig. 60

Fig. 59 illustrates a form of indicator which is attached to one of the forward struts of an airplane at a point where it will be entirely free from any influence of the propeller. AB is a vertical strut on the airplane, to which is attached a wind vane CD, revolving about an axis EF, and actuating a dial K. This wind vane CD is horizontal and always points in the direction of the relative wind, in other words the airplane revolves about the axis EF, as the angle of attack changes. This device is then wired to an electric lamp box indicator similar to the sketch in Fig. 60.

Fastened to the shaft EF (Fig. 59) is a commutator which closes the contact at certain points, and completes the circuit to the different

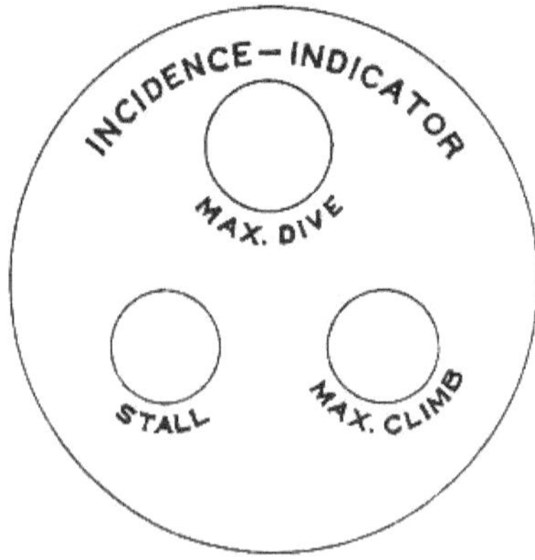

Fig. 60—Indicating dial connected to incidence indicator shown in Fig. 59

Fig. 61—Sperry clinometer

lamps shown in Fig. 60. When the airplane is in danger of stalling the proper light flashes the information to the pilot. The same is true of the other two lights.

The Sperry clinometer is designed with the idea of telling the pilot the inclination of the axis of the airplane. Fig. 61 is an illustration of such an instrument.

Fig. 62 is a cross-section of the same instrument, showing its construction, which, it will be noted, is based on the principle of the plumb-bob. The scale S is mounted on the periphery of a wheel W, which is damped by floating in a liquid. The base of the wheel has a small weight X to maintain it constantly in a vertical condition. There is also an expansion chamber and a filling screw similar to the liquid compass. It is to be noted that this and similar instruments are subject to an error due to lag, caused by the inertia of the wheel for an abrupt change of direction. The clinometer is usually mounted in the cockpit where it can be seen at all times. Skillful pilots pay very little attention to instruments of this kind, however, relying almost entirely on instinct.

The Automatic Drift Set

Before explaining the operation of this instrument, it might be well to include a word in connection with the term "drift," as here used. An aviator, desiring to fly to a point due north of his starting point, with a west wind blowing, will find at the end of a certain length of time that he is considerably to the eastward of his destination, having been blown from his course. To guard against this, two forms of drift indicators are in use. The first is very simple and consists of an eyepiece, containing cross-hairs similar to a surveyor's transit. By looking directly at the ground, the apparent motion of surface objects can be observed and made to coincide with the cross-hairs of the eyepiece. This is done by rotating the eyepiece about the axis of the telescope, and the amount of this rotation, read from a graduated circle, gives the drift angle directly. The pilot then steers "off his course" an amount equal to this angle of drift but in the opposite direction.

A more improved type of this instrument is the Automatic Drift Set shown in Fig. 63. As before, an eyepiece is used to determine the angle of drift as observed from the apparent motion of objects on the ground. As the eyepiece turns to coincide with this line of motion, the connecting cables shown turn the compass case, so that the "lubber line" of the compass is turned automatically in the opposite direction to the drift, and to an equal amount. The pilot then simply steers his predetermined course on the shifted lubber line.

The Bourdon Gauge

This is an instrument for measuring the pressure exerted by a gas or liquid, and is the type used in nearly all steam gauges. It is used on airplanes for registering circulating oil pressure, the pressure of air in

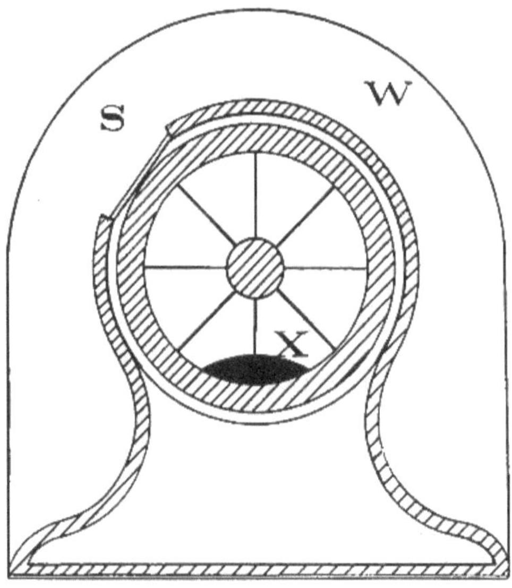

Fig. 62—Cross-section of Sperry clinometer

Fig. 63—Sperry Automatic Drift Set, consisting of a small revolving telescope which is connected to compass case in such a way that compass case with lubber line is automatically turned to compensate for side drift when sighting telescope is turned to line up cross-hairs with apparent motion of plane over ground

the fuel tank, etc. It depends for its action upon the principle that a bent tube if subjected to internal pressure tends to straighten out. Its action can be seen in Fig. 64. A is a tank of air which can be compressed to any desired pressure, as indicated by the pressure gauge C. D is a piece of rubber tube with the outer end closed, and E is a steam gauge from which the dial and covering have been removed. By opening the valve V a pressure is exerted internally in the rubber tube D, tending to straighten it out. The same principle is involved in the metal tube E

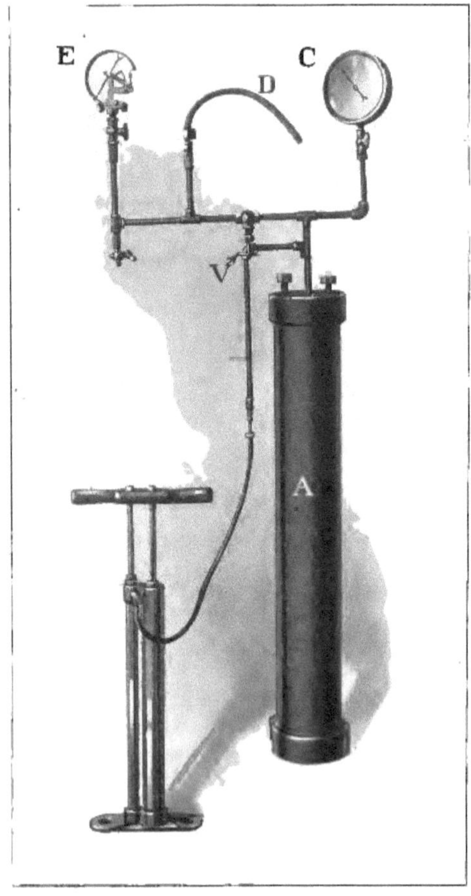

Fig. 64—Test outfit showing principle of operation of Bourdon gauge such as used for indicating steam pressure. This principle is also used in radiator thermometers, air and oil pressure gauges, etc.

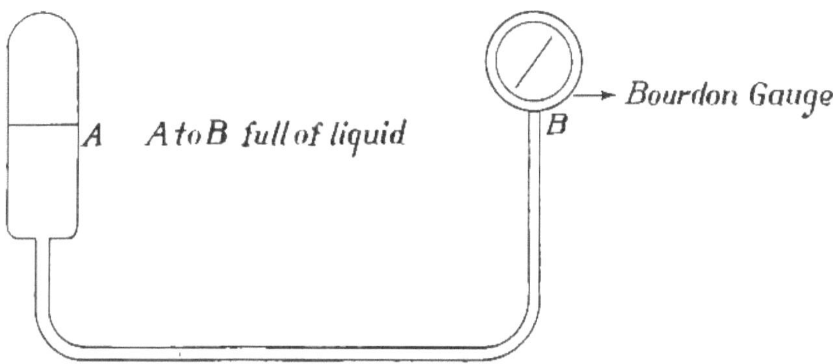

Fig. 65—Showing how a Bourdon gauge, connected to a bulb half full of liquid, is used for a distance thermometer. This device is used in airplanes for indicating radiator temperatures.

which also tends to straighten out. This tube is flattened to increase its sensitiveness, and is connected through a magnifying device to the pointer, which moves in front of a graduated dial.

The Radiator Thermometer

By connecting a Bourdon gauge with a tube of small bore which ends in a chamber or bulb, and then filling the system of gauge and tube with a liquid until the bulb becomes about half full (as shown in Fig. 65) a distance thermometer is obtained, the dial of which can be placed at a point some distance away from the controlling bulb. This type of gauge is used to indicate the temperature of the water in the radiator, so as to guard against overheating or freezing.

As the temperature of the radiator, in which the bulb is placed, rises, more and more of the liquid in the bulb will be changed to vapor, and its pressure will increase accordingly causing the pointer in the Bourdon gauge to move. By properly calibrating the dial the readings can be shown in degrees of temperature in place of pounds of pressure.

Properties of Various Woods

Name of Wood	Weight in pounds per cubic foot (air dried)	Strength in bending		Strength in compression parallel to grain		Stiffness	Hardness	Shock resisting ability	Shearing strength parallel to grain	
		pounds per square inch	percent of oak	pounds per square inch	percent of oak	percent of oak	percent of oak	percent of oak	pounds per square inch	percent of oak
Ash	45	10000	120	4000	125	110	90	125	1500	120
Birch	45	8000	100	3500	100	115	75	125	1000	85
Hickory	50	11000	130	4500	128	120		150	1000	85
Maple	35	8500	100	3500	100	100	75	90	1000	85
Osage Orange	55	13500	150	5500	160	100	190	280		
Walnut	40	9500	110	4000	120	105	85	110	1200	90
Cedar	25	5500	65	3000	85	75	30	45	750	55
Pine	30	6000	70	3000	85	100	30	38	700	50
Spruce	28	6000	70	3000	85	90	35	45	800	60

The above table is mainly comparative. Certain species of these woods may test considerably above or below the figures given.

Chapter 9

NOMENCLATURE FOR AERONAUTICS

Based on official nomenclature recommended by the National Advisory Committee for Aeronautics and definitions used and standardized by the U. S. Army School of Military Aeronautics at Ohio State University.

AERODYNAMICS—The science which treats of the air or other gaseous bodies under the action of forces and of their mechanical effects.

AEROFOIL—A thin wing-like structure, flat or curved, designed to obtain reaction upon its surfaces from the air through which it moves.

AERONAUTICS—That branch of engineering which deals with the design, construction and operation of aircraft.

AILERON—A movable auxiliary surface used for the control of rolling motion of an airplane, i. e., rotation about its fore and aft axis.

AIRCRAFT—Any form of craft designed for the navigation of the air; airplanes, balloons, dirigibles, helicopters, kites, kite balloons, ornithopters, gliders, etc.

AIRDROME—The name usually applied to a ground and buildings used for aviation.

AIRPLANE—A form of aircraft heavier than air, which has wing surfaces for sustentation, stabilizing surfaces, rudders for steering, power plant for propulsion through the air and some form of landing gear; either a gear suitable for rising from or alighting on the ground, or pontoons or floats suitable for alighting on or rising from water. In the latter case, the term "Seaplane" is commonly used. (See definition.)

> **Pusher**—A type of airplane with the propeller or propellers in the rear of the wings.
>
> **Tractor**—A type of airplane with the propeller or propellers in front of the wings.
>
> **Monoplane**—A form of airplane whose main supporting surface is disposed as a single wing extending equally on each side of the body.
>
> **Biplane**—A form of airplane in which the main supporting surface is divided into two parts, one above the other.
>
> **Triplane**—A form of airplane whose main supporting surface is divided into three parts, superimposed.

Multiplane—An airplane the main lifting surface of which consists of numerous surfaces or pairs of superimposed wings.

One and One-Half Plane—A biplane in which the span of the lower plane is decidedly shorter than that of the upper plane.

Flying Boat—An airplane fitted with a boat-like hull suitable for navigation and arising from or alighting on water.

Seaplane—An airplane fitted with pontoons or floats suitable for alighting on or rising from the water.

AIR POCKET—A local movement or condition of the air causing an airplane to drop or lose its correct attitude.

AIR SPEED METER—An instrument designed to measure the velocity of an aircraft with reference to the air through which it is moving.

ALTIMETER—An instrument mounted on an aircraft to continuously indicate its height above the surface of the earth.

ANEMOMETER—An instrument for measuring the velocity of the wind or air currents with reference to the earth or some fixed body.

ANGLE OF ATTACK—The acute angle between the direction of relative wind and the chord of an aerofoil, i. e., the angle between the chord of an aerofoil and its motion relative to the air. (This definition may be extended to any body having an axis.)

Best Climbing—The angle of attack at which an airplane ascends fastest. An angle about half way between the maximum and optimum angle.

Critical—The angle of attack at which the lift is a maximum, or at which the lift curve has its first maximum; sometimes referred to as the "burble point." (If the lift curve has more than one maximum, this refers to the first one.)

Gliding—The angle the flight path makes with the horizontal when flying in still air under the influence of gravity alone, i. e., without power from the engine.

Maximum—The greatest angle of attack at which, for a given power, surface and weight, horizontal flight can be maintained.

Minimum—The smallest angle of attack at which, for a given power, surface and weight, horizontal flight can be maintained.

Optimum—The angle of attack at which the lift-drift ratio is the highest.

ANGLE OF INCIDENCE (Rigger's Angle)—The angle between the longitudinal axis of the airplane and the chord of an aerofoil.

APPENDIX—The hose at the bottom of a balloon used for inflation. In the case of a spherical balloon it also serves for equalization of pressure.

ASPECT RATIO—The ratio of span to chord of an aerofoil.

AVIATOR—The operator or pilot of heavier-than-air craft. This term is applied regardless of the sex of the operator.

AVION—The official French term for military airplanes only.

AXES OF AN AIRCRAFT—The three fixed lines of reference: usually passing through the center of gravity and mutually rectangular. The principal axis in a fore and aft direction, usually parallel to the axis of the propeller and in the plane of symmetry, is the Longitudinal Axis or the Fore-and-Aft Axis.

The axis perpendicular to this and in the plane of symmetry is the Vertical Axis; the third axis perpendicular to the other two is the Lateral Axis, also called the Transverse Axis or the Athwartship Axis. In mathematical discussion the first of these axes, drawn from front to rear is called the X Axis; the second, drawn upward; the Z Axis; and the third, forming a "left-handed" system, the Y Axis.

BALANCED CONTROL SURFACE—A type of surface secured by adding area forward of the axis of rotation. In an airstream a force is exerted on this added area, tending to aid in the movement about the axis.

BALANCING FLAPS—(See **AILERON.**)

BALLONET—A small balloon within the interior of a balloon or dirigible for the purpose of controlling the ascent or descent, and for maintaining pressure on the outer envelope so as to prevent deformation. The ballonet is kept inflated with air at the required pressure, under the control of a blower and valves.

BALLOON—A form of aircraft comprising a gas bag and a basket and supported in the air by the buoyancy of the gas contained in the gas bag, which is lighter than the amount of air it displaces; the form of the gas bag is maintained by the pressure of the contained gas.

Barrage—A small spherical captive balloon, raised as a protection against attacks by airplanes.

Captive—A balloon restrained from free flight by means of a cable attaching it to the earth.

Kite—An elongated form of captive balloon, fitted with tail appendages to keep it headed into the wind, and deriving increased lift due to its axis being inclined to the wind.

Pilot—A small spherical balloon sent up to show the direction of the wind.

Sounding—A small spherical balloon sent aloft, without passengers, but with registering meterological instruments for recording atmospheric conditions at high altitudes.

BALLOON—DIRIGIBLE—A form of balloon the outer envelope of which is of elongated horizontal form, provided with a car, propelling system, rudders and stabilizing surfaces. Dirigibles are divided into three classes: Rigid, Semi-rigid and Non-rigid. In the Rigid type the outer covering is held in place and form by a rigid internal frame work and the shape is maintained independently of the contained gas. The shape and form of the Semi-rigid type is maintained partly by an inner framework and partly by the contained gas. The Non-rigid type is held to form entirely by the pressure of the contained gas.

BALLOON BED—A mooring place on the ground for a captive balloon.

BALLOON CLOTH—The cloth, usually cotton, of which balloon fabrics are made.

BALLOON FABRIC—The finished material, usually rubberized, of which balloon envelopes are made.

BANK—To incline an airplane laterally, i. e., to rotate it about the fore-and-aft axis when making a turn. Right bank is to incline the airplane with the right wing down. Also used as a noun to describe the position of an airplane when its lateral axis is inclined to the horizontal.

BAROGRAPH—An instrument for recording variations in barometric pressure. In aeronautics the charts on which the records are made are prepared to indicate altitudes directly instead of barometric pressure, inasmuch as the atmospheric pressure varies almost directly with the altitude.

BAROMETER—An instrument for measuring the pressure of the atmosphere.

BASKET—The car suspended beneath the balloon for passengers, ballast, etc.

BIPLANE—(See **AIRPLANE.**)

BODY (OF AN AIRPLANE)—A structure, usually enclosed, which contains in a streamline housing the power plant, fuel, passengers, etc.

> **Fuselage**—A type of body of streamline shape carrying the empannage and usually forming the main structural unit of an airplane.
>
> **Monocoque**—A special type of fuselage constructed of metal sheeting or laminated wood. A monocoque is generally of circular or elliptical cross-section.
>
> **Nacelle**—A type of body shorter than a fuselage. It does not carry the empannage, but acts more as a streamline housing. Usually used on a pusher type of machine.
>
> **Hull**—A boat-like structure which forms the body of a flying-boat.

BONNET—The appliance, having the form of a parasol, which protects the valve of a spherical balloon against rain.

BOOM—(See **OUTRIGGER.**)

BOWDEN WIRE—A stiff control wire enclosed in a tube used for light control work where the strain is comparatively light, as for instance throttle and spark controls, etc.

BOWDEN WIRE GUIDE—A close wound, spring-like, flexible guide for Bowden wire controls.

BRIDLE—The system of attachment of cables to a balloon, including lines to the suspension band.

BULLS EYES—Small rings of wood, metal, etc., forming part of balloon rigging, used for connection or adjustment of ropes.

BURBLE POINT—(See **ANGLE—CRITICAL.**)

CABANE (OR CABANE STRUT)—In a monoplane, the strut or pyramidal frame work projecting above the body and wings and to which the stays, ground wires, braces, etc., for the wing are attached.

In a biplane, the compression member of an auxiliary truss, serving to support the overhang of the upper wing.

CAMBER—The convexity or rise of the curve of an aerofoil from its chord, usually expressed as the ratio of the maximum departure of the curve from the chord as a fraction thereof. **Top Camber** refers to the top surface and **Bottom Camber** to the bottom surface of an aerofoil. **Mean Camber** is the mean of these two.

CAPACITY-CARRYING—The excess of the total lifting capacity over the dead load of an aircraft. The latter includes structure, power plant and essential accessories. Gasoline and oil are not considered essential accessories.

The cubic contents of a balloon.

CAPACITY-LIFTING—(See **LOAD**)—The maximum flying load of an aircraft.

CATHEDRAL—A negative dihedral.

CEILING—The maximum possible altitude to which a given airplane can climb.

CENTER—The point in which a set of effects is assumed to be accumulated, producing the same effect as if all were centered at this point.

There are five main centers in an airplane—Center of Lift, Center of Gravity, Center of Thrust, Center of Drag and Center of Keelplane Area. The latter is also called the Directional Center. The stability, controllability and general air worthiness of an airplane depend largely on the proper positioning of these centers.

CENTER OF PRESSURE OF AN AEROFOIL—The point in the plane of the chords of an aerofoil, prolonged if necessary, through which at any given attitude the line of action of the resultant air force passes. (This definition may be extended to any body.)

CENTER PANEL—The central part of the upper wing (of a biplane) above the fuselage. The upper wings are attached to this on either side.

CHORD—(Of an aerofoil section.) A straight line tangent to the under curve of the aerofoil section, front and rear.

CHORD LENGTH—(Or length of Chord.)—The length of an aerofoil section projected on the chord, extended if necessary.

CLINOMETER—(See **INCLINOMETER**.)

CLOCHE—The bell-shaped construction which forms the lower part of the pilot's control lever in the Bleriot control and to which the control cables are attached.

COCKPIT—The space in an aircraft body occupied by pilots or passengers.

CONCENTRATION RING—The hoop to which are attached the ropes suspending the basket (of a balloon).

CONTROLS—A general term applied to the mechanism used to control the speed, direction of flight and altitude of an aircraft.

Bridge (Deperdussin-"Dep" Control)—An inverted "U" frame pivoted near its lower points, by which the motion of the elevators is controlled. The ailerons are controlled by a wheel mounted on the upper center of this bridge.

Dual—Two sets of inter-connected controls allowing the machine to be operated by one or two pilots.

Shoulder—A yoke fitting around the shoulders of the pilot by means of which the ailerons are operated (by the natural side movement of the pilot's body) to cause the proper amount of banking when making a turn or to correct excessive bank. (Used on early Curtiss planes.)

Stick (Joy-stick)—A vertical lever pivoted near its lower end and used to operate the elevators and ailerons.

COWLS—The metal covering enclosing the engine section of the fuselage.

CROW'S FOOT—A system of diverging short ropes for distributing the pull of a single rope. (Used principally on balloon nets.)

DECALAGE—The difference in the angular setting of the chord of the upper wing of a biplane with reference to the chord of the lower wing.

DIHEDRAL (In an airplane)—The angle included at the intersection of the imaginary surfaces containing the chords of the right and left wings (continued to the plane of symmetry if necessary). This angle is measured in a plane perpendicular to that intersection. The measure of the dihedral is taken as 90 deg. minus one-half of this angle as defined.

The dihedral of the upper wing may and frequently does differ from that of the lower wing in a biplane.

> **Lateral**—An airplane is said to have lateral dihedral when the wings slope downward from the tips toward the fuselage.
>
> **Longitudinal**—The angular difference between the angle of incidence of the main planes and the angle of incidence of the horizontal stabilizer.

DIRIGIBLE—A form of balloon, the outer envelope of which is of elongated horizontal form, provided with a propelling system, car, rudders and stabilizing surfaces.

> **Non-Rigid**—A dirigible whose form is maintained by the pressure of the contained gas assisted by the car suspension system.
>
> **Rigid**—A dirigible whose form is maintained by a rigid structure contained within the envelope.
>
> **Semi-Rigid**—A dirigible whose form is maintained by means of a rigid keel and by gas pressure.

DIVING RUDDER—(See **ELEVATOR.**)

DOPE—A preparation, the base of which is cellulose acetate or cellulose nitrate, used for treating the cloth surfaces of airplane members or the fabric of balloon gas bags. It increases the strength of the fabric, produces tautness, and acts as a filler to make the fabric impervious to air and moisture.

DRAG—The component parallel to the relative wind of the total force on an aircraft due to the air through which it moves.

That part of the drag due to the wings is called "Wing Resistance" (formerly called "Drift"); that due to the rest of the airplane is called "Parasite Resistance" (formerly called head resistance).

The total resistance to motion through the air of an aircraft, that is, the sum of the drift and parasite resistance. Total Resistance.

DRIFT—The component of the resultant wind pressure on an aerofoil or wing surface parallel to the air stream attacking the surface.

Also used as synonymous with lee-way.

(See **DRAG.**)

DRIFT INDICATOR—An instrument for the measurement of the angular deviation of an aircraft from a set course, due to cross winds.

Also called **Drift Meter.**

DRIFT WIRES—Wires which take the drift load and transfer it through various members to the body of the airplane.

DRIP CLOTH—A curtain around the equator of a balloon which prevents rain from dripping into the basket.

DROOP—

(a) An aileron is said to have droop when it is so adjusted that its trailing edge is below the trailing edge of the main plane.

(b) When a wing is warped to give wash-out or wash-in, its trailing edge will, relative to the leading edge, be displaced progressively from one end to the other. A downward displacement is called droop.

ELEVATOR—A hinged surface, usually in the form of a horizontal rudder, mounted at the tail of an aircraft for controlling the longitudinal attitude of the aircraft, i. e., its rotation about the lateral axis.

EMPANNAGE—A term applied to the tail group of parts of an airplane. (See **TAIL.**)

ENGINE SILL, BEARERS, SUPPORTS—The members forming the engine bed.

ENTERING EDGE—The foremost part or forward edge of an aerofoil or propeller blade.

ENVELOPE—The portion of the balloon or dirigible which contains the gas.

EQUATOR—The largest horizontal circle of a spherical balloon.

FAIRING—A wood or metal form attached to the rear of struts, braces or wires to give them a streamline shape.

FAIR LEAD—A guide for a cable.

FIN—A small fixed aerofoil attached to part of an aircraft to promote stability; for example, tail fin, skid fin, etc. Fins may be either horizontal or vertical and are often adjustable.

(See **STABILIZER.**)

FIRE DASH—A metal screen dividing the engine section of an airplane body from the cockpit section.

FLIGHT PATH—The path of the center of gravity of an aircraft with reference to the earth.

FLOAT—That portion of the landing gear of an aircraft which provides buoyancy when it is resting on the surface of the water.

FLYING BOAT—(See **AIRPLANE.**)

FLYING POSITION—The position of a machine, assumed when flying horizontally in still air. When on the ground the machine is placed in a flying position by leveling both longitudinally and laterally. The two longerons, engine sills or other perpendicular parts designated by the maker are taken as reference points from which to level.

FOOT BAR—(See **RUDDER BAR.**)

FUSELAGE—(See **BODY.**)

FUSELAGE COVER—A cover placed on a fuselage to preserve a streamline shape.

GAP—The shortest distance between the planes of the chords of the upper and lower wings of a biplane.

GAS BAG—(See **ENVELOPE.**)

GLIDE—To fly without power and under the influence of gravity alone.

GLIDER—A form of aircraft similar to an airplane but without any power plant.

When utilized in variable winds it makes use of the soaring principles of flight and is sometimes called a soaring machine.

GLIDING ANGLE—(See **ANGLE.**)

GORE—One of the segments of fabric comprising the envelope of a balloon.

GROUND CLOTH—Canvas placed on the ground to protect a balloon.

GUIDE ROPE—A long trailing rope attached to a spherical balloon or dirigible to serve as a brake and as a variable ballast.

GUY—A rope, chain, wire or rod attached to an object to guide or steady it, such as guys to wing, tail or landing gear.

HANGAR—An airplane shed.

HEAD RESISTANCE—(See **PARASITE RESISTANCE.**)

HELICOPTER—A form of aircraft whose support in the air is derived from the vertical thrust of propellers.

HORN-CONTROL ARM—An arm at right angles to a control surface to which a control cable is attached, for example, aileron horn, rudder horn, elevator horn, etc. More commonly called a **Mast.**

HULL—(See **BODY.**)

INCLINOMETER—An instrument for measuring the angle made by the axis of an aircraft with the horizontal.

Indicator-Banking—An inclinometer indicating lateral inclination or bank.

INSPECTION WINDOW—A small transparent window in the envelope of a balloon or in the wing of an airplane to allow inspection of the interior, or of aileron controls when the latter are mounted inside an aerofoil section.

INSTABILITY—An inherent condition of a body, which, if the body is distributed, causes it to move toward a position away from its first position, instead of returning to a condition of equilibrium.

KEEL PLANE AREA—The total effective area of an aircraft which acts to prevent skidding or side slipping.

KITE—A form of aircraft without other propelling means than the tow-line pull, whose support is derived from the force of the wind moving past its surfaces.

LANDING GEAR—The understructure of an aircraft designed to carry the load when resting on, or running on, the surface of the land or water.

LEADING EDGE—(See **ENTERING EDGE.**)

LEEWAY—The angle of deviation from a set course over the earth, due to cross currents of wind. Also called Drift.

LIFT—The component of the force due to the air pressure of an aerofoil resolved perpendicular to the flight path in a vertical plane.

LIFT-DRIFT RATIO—The proportion of lift to drift is known as the lift-drift ratio. It expresses the efficiency of the aerofoil.

LIFT BRACING—(See **STAY.**)

LOAD—

Dead—The structure, power plant and essential accessories of an aircraft.

Full—The maximum weight which an aircraft can support in flight; the gross weight.

Useful—The excess of the full load over the dead weight of the aircraft itself, i. e., over the weight of its structure, power plant and essential accessories. (These last must be specified.)

(See **Capacity.**)

LOADING—The weight carried by an aerofoil, usually expressed in pounds per square foot of superficial area.

LOBES—Bags at the stern of an elongated balloon designed to give it directional stability.

LONGERON—The principal fore-and-aft structural members of the fuselage or nacelle of an airplane.

(See **LONGITUDINAL.**)

LONGITUDINAL—A fore-and-aft member of the framing of an airplane body, or of the float in a seaplane, usually continuous across a number of points of support.

LONGITUDINAL DIHEDRAL—(See **DIHEDRAL.**)

MAST—(See **HORN.**)

MONOCOQUE—(See **BODY.**)

MONOPLANE—A form of airplane whose main supporting surface is a single wing extending equally on each side of the body.

(See **AIRPLANE.**)

MOORING BAND—The band of tape over the top of a balloon to which are attached the mooring ropes.

NACELLE—(See **BODY.**)

NET—A rigging made of ropes and twine on spherical balloons, which supports the entire load carried.

NOSE DIVE—A dangerously steep descent, head on.

NOSE PLATE—A plate at the nose or front end of the fuselage in which the longerons terminate.

NOSE SPIN—A nose dive in which the airplane rotates about its own axis due to the reaction from the propeller. It usually results from failure to shut off the engine in time when going into a nose dive, and is likely to cause complete loss of control.

ORNITHOPTER—A form of aircraft deriving its support and propelling force from flapping wings.

OUT-RIGGER—Members, independent of the body, extending forward or to the rear and supporting control or stabilizing surfaces.

OVERHANG—The distance the wings project out beyond the outer struts.

PAN CAKE, TO—To descend as a parachute after a machine has lost forward velocity. To strike the ground violently without much forward motion.

PANEL—A portion of a framed structure between adjacent posts or struts. Applied to the fuselage it is the area bounded by two struts and the longerons. An entire wing is often spoken of as a panel. Thus the upper lifting surface of a biplane is usually of three parts designated as the right upper panel, left upper panel and the center panel.

PARACHUTE—An apparatus made like an umbrella used to retard the descent of a falling body.

PARASITE RESISTANCE—The total resistance to motion through the air of all parts of an aircraft not a part of the main lifting surface.

PATCH SYSTEM—A system of construction in which patches or adhesive flaps are used in place of the suspension band in a balloon.

PERMEABILITY—The measure of the loss of gas by diffusion through the intact balloon fabric.

PHILLIPS ENTRY—A reverse curve on the lower surface of an aerofoil, towards the entering edge, designed to more evenly divide the air.

PITCH-OF A PROPELLER—(See **PROPELLER.**)

PITCH-OF A SCREW—The distance a screw advances in its nut in one revolution.

PITCH, TO—To plunge in a fore-and-aft direction.

PITOT TUBE—A tube with an end open square to the fluid stream, used as a detector of an impact pressure. It is usually associated with a concentric tube surrounding it, having perforations normal to the axis for indicating static pressure; or there is such a tube placed near it and parallel to it, with a closed conical end and having perforations in its side. The velocity of the fluid can be determined from the difference between the impact pressure and the static pressure, as read by a suitable gauge. This instrument is often used to determine the velocity of an aircraft through the air.

PLANE OF SYMMETRY—A vertical plane through the longitudinal axis of an airplane. It divides the airplane into two symmetrical portions.

PONTOON—(See **FLOAT.**)

PROPELLER OR AIR SCREW—A body so shaped that its rotation about an axis produces a thrust in the direction of its axis.

Disc-Area of Propeller—The total area of a circle swept by the propeller tips.

Pitch Of—The distance a propeller will advance in one revolution, supposing the air to be solid.

Race—The stream of air driven aft by the propeller and with a velocity relative to the airplane greater than that of the surrounding body of still air. (Frequently called slip-stream.)

Slip Of—The difference between the distance a propeller actually advances and the distance it would advance while making the same number of revolutions in a solid medium. Usually expressed as a percentage of the total distance.

Torque Of—The turning moment of the propeller. The effect of propeller torque is an equal reaction tending to rotate the whole airplane in the opposite direction to that of the propeller.

PUSHER—(See **AIRPLANE.**)

PYLON—A post, mast or pillar serving as a marker of a flying course. Also used infrequently to designate the control masts such as the aileron mast, rudder mast, elevator mast, etc.

RAKE—The angular deviation of the outer end of a wing from a line at right angles to the entering edge.

RELATIVE WIND—The motion of the air with reference to a moving body. Its direction and velocity, therefore, are found by adding two vectors, one being the velocity of the air with reference to the earth, the other being equal and opposite to the velocity of the body with reference to the earth.

RETREAT—(See **SWEEP BACK.**)

RIB—A member used to give strength and shape to an aerofoil in a fore-and-aft direction.

> **Web**—A light rib, the central part of which is cut out in order to lighten it.
>
> **Compression**—A rib heavier than the web type and so constructed as to resist the compression due to the wire bracing of the airplane.
>
> **Secondary Nose**—Small ribs extending from the front spar to the nose strip (entering edge). Placed between the main ribs to give support to the fabric near the entering edge. Sometimes called Stub Ribs.

RIGGING—The art of truing up an airplane and keeping it in flying condition.

RIP CORD—The rope running from the rip panel of a balloon to the basket, the pulling of which causes immediate deflation.

RIP PANEL—A strip in the upper part of a balloon which is torn off when immediate deflation is desired.

RUDDER—A hinged or pivoted surface, usually more or less flat or streamlined, used for the purpose of controlling the attitude of an aircraft about its vertical axis, i. e., for controlling its lateral movement.

RUDDER BAR—A bar pivoted at the center, to the ends of which the rudder control cables are attached. The pilot operates the rudder by moving the rudder bar with his feet.

RUDDER POST—The post to which the rudder is hinged, generally forming the rear vertical member of the vertical stabilizer.

SEA PLANE—An airplane fitted with pontoons or floats suitable for alighting on or rising from the water.

(See **AIRPLANE.**)

SERPENT—A short heavy guide rope used with balloons.

SERVING—A binding of wire, cord or other material. Usually used in connection with joints in wood, and cable splices.

SIDE SLIPPING—Sliding sideways and downward toward the center of a turn, due to an excessive amount of bank. It is the opposite of skidding.

SIDE WALK—A reinforced portion of the wings near the fuselage serving as a support in climbing about the airplane. Otherwise known as running board.

SKIDDING—Sliding sideways away from the center of a turn, due to an insufficient amount of bank. It is the opposite of side slipping.

SKIDS-LANDING GEAR—Long wooden or metal runners designed to prevent nosing of a land machine when landing, or to prevent dropping into holes or ditches in rough ground. Generally designed to function in case the wheels should collapse or fail to act.

Tail—A skid supporting the tail of a fuselage while on the ground.

Wing—A light skid placed under the lower wing to prevent possible damage on landing.

SKIN FRICTION—Friction between the air and a surface over which it is passing

SLIP STREAM—(See **PROPELLER RACE.**)

SOARING MACHINE—(See **GLIDER.**)

SPAN-WING—Span is the dimension of a surface across the air stream.

Wing Span or Spread of a machine is length overall from tip to tip of wings.

SPARS-WING—Long pieces of wood or other material forming the main supporting members of the wing, and to which the ribs are attached.

SPREAD—(See **SPAN.**)

STABILITY—The quality of an aircraft in flight which causes it to return to a condition of equilibrium after meeting a disturbance.

Directional—That property of an airplane by virtue of which it tends to hold a straight course. That is, if a machine tends constantly to veer off its course necessitating exercise of the controls by the pilot to keep it on its course, it is said to lack directional stability.

Dynamical—The quality of an aircraft in flight which causes it to return to a condition of equilibrium after its attitude has been changed by meeting some disturbance, e. g., a gust. This return to equilibrium is due to two factors; first, the inherent righting moments of the structure; second, the damping of the oscillations by the tail, etc.

Inherent—Stability of an aircraft due to the disposition and arrangement of its fixed parts, i. e., that property which causes it to return to its normal attitude of flight without the use of the controls.

Lateral—The property of an airplane by virtue of which the lateral axis tends to return to a horizontal position after meeting a disturbance.

Longitudinal—An airplane is longitudinally stable when it tends to fly on an even keel without pitching or plunging.

Statical—In wind tunnel experiments it is found that there is a definite angle of attack such that for a greater angle or a less one the righting moments are in such a sense as to tend to make the attitude return to this angle. This holds true for a certain range of angles on each side of this definite angle; and the machine is said to possess "statical stability" through this range.

STABILIZER—Balancing planes of an aircraft to promote stability.

Horizontal—A horizontal fixed plane in the empennage designed to give stability about the lateral axis.

Vertical—A vertical fixed plane in the empennage to promote stability about the vertical axis.

Mechanical—Any mechanical device designed to secure stability in flight.

STABILIZING FINS—Vertical surfaces mounted longitudinally between planes, to increase the keel plane area.

STAGGER—The amount of advance of the entering edge of a superposed aerofoil of an airplane, over that of a lower, expressed as a percentage of the gap. It is considered positive when the upper aerofoil is forward.

STALLING—A term describing the condition of an airplane which, from any cause has lost the relative speed necessary for steerageway and control.

STATION—The points at which struts join the longerons in a fuselage, are termed stations and are numbered according to some arbitrary system. Some makers begin with No. 1 at the nose plate and number toward the rear. Other makers begin with 0 at the tail post and number toward the front.

STATOSCOPE—An instrument to detect the existence of a small rate of ascent or descent, principally used in ballooning.

STAY—A wire, rope, or the like used as a tie piece to hold parts together, or to contribute stiffness; for example, the stays of the wing and body trussing.

STREAMLINE-FLOW—A term used to describe the condition of continuous flow of a fluid, as distinguished from eddying flow, where discontinuity takes place.

STREAMLINE-SHAPE—A shape intended to avoid eddying or discontinuity and to preserve streamline-flow, thus keeping resistance to progress at a minimum.

STRINGERS—A term applied to the slender wooden members running laterally through the wing ribs for the purpose of stiffening them.

STRUT—A compression member of a truss frame, for instance, the vertical members of the wing truss of a biplane.

STRUT-INTERPLANE—A strut holding apart two aerofoils.

SUPPORTING SURFACE—Any surface of an airplane on which the air produces a lift reaction.

SUSPENSION BAND—The band around a balloon to which are attached the basket and the main bridle suspensions.

SUSPENSION BAR—The bar used for the concentration of basket suspension ropes in captive balloons.

SWEEP-BACK—The horizontal angle between the lateral (athwartship) axis of an airplane and the entering edge of the main planes.

TACHOMETER—An instrument for indicating the number of revolutions per minute of the engine or propeller.

TAIL CUPS—The steadying device attached at the rear of certain types of elongated captive balloons.

TAIL-NEUTRAL—A tail, the horizontal stabilizer of which is so set that it gives neither an upward lift nor a downward thrust when the machine is in normal flight.

Positive—A tail in which the horizontal stabilizer is so set as to give an upward lift and thus assist in carrying the weight of the airplane when it is in normal flight.

Negative—One in which the horizontal stabilizer is so set as to give a downward thrust on the tail when the machine is in normal flight.

TAIL POST—The vertical strut at the rear end of the fuselage.

TAIL SKID—A skid supporting the tail of a fuselage while on the ground.

TAIL SLIDE—A steep descent, tail downward. Usually caused by stalling on an attempt to climb too steeply.

THIMBLE—An elongated metal eye spliced in the end of a rope or cable.

TRACTOR—(See **AIRPLANE**.)

TRAILING EDGE—The rearmost portion of an aerofoil.

TRIPLANE—A form of airplane whose main supporting surface is divided into three parts, superimposed.

TRUSS—The framing by which the wing loads are transmitted to the body; comprises struts, stays and spars.

UNDERCARRIAGE—(See **LANDING GEAR**.)

VETTING—The process of sighting by eye along edges of spars, planes, etc., to ascertain their alignment. An experienced man can detect and remedy many faults in alignment by this method.

VOL-PIQUE'—(See **NOSE DIVE**.)

VOLPLANE—To glide.

WARP—To change the form of the wing by twisting it, usually by changing the inclination of the rear spar relative to the front spar.

WASHIN—A progressive increase in the angle of incidence from the fuselage toward the wing tip.

WASHOUT—A progressive decrease in the angle of incidence from the fuselage toward the wing tip.

WEIGHT-GROSS—(See **LOAD, FULL**.)

WINGS—The main supporting surfaces of an airplane. Also called Aerofoils.

WING FLAPS—(See **AILERON**.)

WING LOADING—(See **LOADING**.)

WING MAST—The mast structure projecting above the wing, to which the top load wires are attached.

WING RIB—A fore-and-aft member of the wing structure used to support the covering and to give the wing section its form. (See **RIB**.)

WING SPAR OR WING BEAM—A transverse member of the wing structure. (See **SPARS-WING**.)

WIRES—

Drift—Wires that take the drift load and transfer it through various members to the body of the airplane.

Flying—The wires that transfer to the fuselage, the forces due to the lift on the wings when an airplane is in flight. They prevent the wings from collapsing upwards during flight.

Landing—The wires that transfer to the fuselage, the forces due to the weight of the wings when an airplane is landing or resting on the ground.

Stagger—The cross brace wires between the interplane struts in a fore-and-aft direction.

YAW—To yaw is to swing off the course and turn about the vertical axis owing to side gusts of wind or lack of directional stability.

Angle Of—The temporary angular deviation of the fore-and-aft axis from the course.

Physical and Mechanical Terms

ACCELERATION—The rate of increase of velocity.

CENTER OF GRAVITY—The center of gravity of a body is that point about which, if suspended, all the parts will be in equilibrium, that is, there will be no tendency to rotation.

CENTRIFUGAL FORCE—That force which urges a body, moving in a curved path, outward from the center of rotation.

COMPONENT—A force which when combined with one or more like forces produces the effect of a single force. The single force is regarded as the RESULTANT of the component forces.

DENSITY—Mass per unit of volume; for instance, pounds per cubic foot.

EFFICIENCY—(Of a machine.)—The ratio of output to input of power, usually expressed as percentage.

ELASTIC LIMIT—The greatest stress per unit area which will not produce a permanent deformation of the material under stress.

ELONGATION—When any material fails by tension it usually stretches and takes a permanent set before it breaks. The ratio of this permanent elongation to the original length, expressed as a percentage, is a measure of the elongation.

ENERGY—The capacity of a body for doing work. Heat is a form of energy. Any chemical reaction that generates heat or electricity liberates energy. Bodies may possess energy by virtue of having work done upon them.

EQUILIBRIUM—When two or more forces act upon a body in such a way that no motion results, there is said to be equilibrium.

FACTOR OF SAFETY—The ratio of the load required to cause failure in a structural member to the usual working load the member is designed to carry. Thus if a member be designed to carry a load of 500 lbs. and it would require a load of 2000 lbs. to cause failure, the factor of safety would be four.

FOOT-POUND—The foot-pound is a unit of work. It is equal to a force of one pound acting through a distance of one foot. This is a foot-pound of energy.

INERTIA—That property of a body by virtue of which it resists any attempt to start it if at rest, to stop it if in motion, or in any way to change either the direction or velocity of motion, is called **Inertia.**

MASS—The mass of a body is a measure of the quantity of material in it.

MOMENT—Moment is the product of a force times its lever arm. It is usually expressed in **Inch-Pounds.**

MOMENTUM—Momentum is the product of the mass and velocity of a moving body. It is a measure of the quantity of motion.

POWER—Power is the time rate of doing work.

Horsepower—The horsepower is a unit of work. One horsepower represents the performance of work at the rate of 33,000 foot-pounds per minute, or 550 foot-pounds per second.

RESULTANT OF A FORCE—The resultant of two or more forces is that single force which will produce the same effect upon a body as is produced by the joint action of the component forces.

STRESS—The internal condition of a body under the action of opposing forces. The unit of measure is usually pounds per square inch.

Compression—When forces are applied to a body in such a way as to tend to crush it, there results a compressive stress in the body.

Tension—When forces are applied to a body in such a way as to tend to separate or pull it apart, the body is said to be in tension or a tensile stress has been produced within it.

Shear—When external forces are applied in such a way as to cause a tendency for particles of a body to slip or slide past each other, there results a shearing stress in the body.

STRAIN—Strain is the deformation produced in a body by the application of external forces.

TORQUE—When forces are so disposed as to cause or tend to cause rotation, there is produced a turning moment which is also called torque. It is usually measured in inch-pounds. Thus if a force of 10 pounds be applied tangentially to the rim of a wheel of 10-inch radius, the torque or turning moment will be 100 inch-pounds.

ULTIMATE STRENGTH—The load per square inch required to produce fracture.

VELOCITY—In uniform motion, the distance passed over in a unit of time, as one second. This may also be obtained by dividing the length of any portion of the path by the time taken to describe that portion, no matter how small or great.

In variable motion, where velocity varies from point to point, its value at any point is expressed as the quotient of an infinitely small distance, containing the given point by the infinitely small portion of time in which this distance is described.

WORK—The product of a force by the distance described in the direction of the force by the point of application. If the force moves forward it is called a working force, and is said to do the work expressed by this product; if backward, it is called a resistance, and is then said to have the work done upon it, in overcoming the resistance through the distance mentioned (it might also be said to have done negative work).

In a uniform translation, the working forces do an amount of work which is entirely applied to overcoming the resistances.

Zeitfracht Medien GmbH
Ferdinand-Jühlke-Straße 7
99095 Erfurt, Deutschland
produktsicherheit@kolibri360.de